Rees
Understanding Injection Mold Design

Hanser **Understanding** Books
A Series of Mini-Tutorials

Series Editor: E.H. Immergut

Understanding Injection Molding Technology (Rees)
Understanding Polymer Morphology (Woodward)
Understanding Product Design for Injection Molding (Rees)
Understanding Design of Experiments (Del Vecchio)
Understanding Plastics Packaging Technology (Selke)
Understanding Compounding (Wildi/Maier)
Understanding Extrusion (Rauwendaal)
Understanding Thermoforming (Throne)
Understanding Thermoplastic Elastomers (Holden)
Understanding Blow Molding (Lee)
Understanding Injection Mold Design (Rees)

Herbert Rees

Understanding Injection Mold Design

HANSER

Hanser Publishers, Munich

Hanser Gardner Publications, Inc., Cincinnati

The Author:
Herbert Rees, 248386-5 Sideroad (moro), RR#5 Orangeville, Ontario, Canada, L9W 2Z2

Distributed in the USA and in Canada by
Hanser Gardner Publications, Inc.
6915 Valley Avenue, Cincinnati, Ohio 45244-3029, USA
Fax: (513) 527-8950
Phone: (513) 527-8977 or 1-800-950-8977
Internet: http://www.hansergardner.com

Distributed in all other countries by
Carl Hanser Verlag
Postfach 86 04 20, 81631 München, Germany
Fax: +49 (89) 98 12 64
Internet: http://www.hanser.de

The use of general descriptive names, trademarks, etc., in this publication, even if the former are not especially identified, is not to be taken as a sign that such names, as understood by the Trade Marks and Merchandise Marks Act, may accordingly be used freely by anyone.

While the advice and information in this book are believed to be true and accurate at the date of going to press, neither the authors nor the editors nor the publisher can accept any legal responsibility for any errors or omissions that may be made. The publisher makes no warranty, express or implied, with respect to the material contained herein.

Library of Congress Cataloging-in-Publication Data

Rees, Herbert, 1915–
    Understanding injection mold design/Herbert Rees.
        p. cm.
    Includes bibliographical references and index.
    ISBN 1-56990-311-5 (softback)
        1. Injection molding of plastics. I. Title.

    TP1150.R45 2001
    668.4'12–dc21   00-054085

Die Deutsche Bibliothek - CIP-Einheitsaufnahme

Rees, Herbert:
Understanding injection mold design/Herbert Rees, -Munich : Hanser;
Cincinnati:Hanser Gardner, 2001
    (Hanser understanding books)
    ISBN 3-446-21587-5

© Carl Hanser Verlag, Munich 2001
Typeset in the U.K. by Techset Composition Ltd., Salisbury
Printed and bound in Germany by Druckhaus "Thomas Müntzer", Bad Langensalza

Introduction to the Series

In order to keep up in today's world of rapidly changing technology we need to open our eyes and ears and, most importantly, our minds to new scientific ideas and methods, new engineering approaches and manufacturing technologies and new product design and applications. As students graduate from college and either pursue academic polymer research or start their careers in the plastics industry, they are exposed to problems, materials, instruments and machines that are unfamiliar to them. Similarly, many working scientists and engineers who change jobs must quickly get up to speed in their new environment.

To satisfy the needs of these "newcomers" to various fields of polymer science and plastics engineering, we have invited a number of scientists and engineers, who are experts in their field and also good communicators, to write short, introductory books which let the reader "understand" the topic rather than to overwhelm him/her with a mass of facts and data. We have encouraged our authors to write the kind of book that can be read profitably by a beginner, such as a new company employee or a student, but also by someone familiar with the subject, who will gain new insights and a new perspective.

Over the years this series of **Understanding** books will provide a library of mini-tutorials on a variety of fundamental as well as technical subjects. Each book will serve as a rapid entry point or "short course" to a particular subject and we sincerely hope that the readers will reap immediate benefits when applying this knowledge to their research or work-related problems.

E.H. Immergut
Series Editor

Preface

During the last fifty years I have been almost continuously working with molders, mold makers and mold designers, and in doing so learning the intricacies of designing of molds for many different products, from the early, simple compression molds to highly sophisticated injection molds. I have worked with them not only in North America, but also in Europe and Japan, and especially in the last 15 years, as consultant to those in developing countries who only recently started to seriously compete in the huge field of manufacturing molded plastic products.

During my discussions with these newcomers to the field, but also in earlier years, when talking to "old hands" in this field, I have often wondered how many of them really understood what they were doing when it comes to planning for and designing a new mold, and why they were doing it. In many cases I believe they took simply "the easy way out" by just imitating what they saw in other molds, and expanding on it, regardless of whether the molds used as "precedents" were for comparable conditions, for the same plastic, for similar molding machines, or for a similar production requirement. Another problem I saw was that in many mold making shops, here and everywhere, some designers were more intent on making "pretty pictures", in the shortest posssible time, rather than understanding that the job expected of a mold designer is to consider possible alternatives of how the planned mold could look, then make a practical and most suitable layout of a mold to produce the best quality product, at the lowest cost, and finally supply all pertinent information to the mold maker, the machinists, and asssemblers.

With the advent of computer aided designing (CAD), the technique of making mold designs and drawings has become much easier to handle, and in some cases where products are similar, it has become often so simple that the mold design can be performed almost automatically, by just following the prompts of the computer, by recalling older complete or partial designs from the CAD memory, and creating a new mold by just changing some dimensions. If you are brought up in this environment, you may be able to produce good designs, based on the available good precedents, but you will be hard pressed to generate a good mold for which there is no precedent on file.

I undertook to write this book "Understanding Injection Mold Design" essentially to explain what is really important in the design of an injection mold, so that a good mold, best suitable for the application, can be created even if there is no precedent. It is meant to be used to guide the designer to think, and to frequently ask why, where, when, how, etc., when considering the many possible choices before settling on a final concept. Also, in my experience, the greatest obstacle to creating a good design has always been the reluctance of the designer to acknowledge the possibility that he or she may be wrong, and that there may be a better way than the first one proposed. The designer must never forget, it is always cheaper to change a design layout even if it adds some design time, than to change (re-machine or modify) a poorly designed but already built mold.

Herbert Rees,
Orangeville, ON

Contents

1 Introduction 1
   1.1 Economics of Mold Design 3

2 Starting New in the Mold Design Field 5

3 The Basics of an Injection Molding Machine 7

4 Understanding the Basics of the Injection Mold 9
   4.1 Design Rules 9
   4.2 The Basic Mold 9
       4.2.1 Mold Cavity Space 9
       4.2.2 Number of Cavities 10
       4.2.3 Cavity Shape and Shrinkage 11
   4.3 Cavity and Core 11
   4.4 The Parting Line 12
       4.4.1 Split Molds and Side Cores 13
   4.5 Runners and Gates 14
   4.6 Projected Area and Injection Pressure 15
       4.6.1 Clamping Force 16
       4.6.2 Strength of the Mold 16
       4.6.3 Why are High Injection Pressures Needed? 17
   4.7 Venting 18
   4.8 Cooling 18
       4.8.1 Basics of Cooling 19
       4.8.2 Plate Cooling 22
   4.9 Ejection 22
       4.9.1 Automatic Molding 25
   4.10 Shrinkage 26
       4.10.1 Variable Shrinkage 26
   4.11 Alignment 27

4.11.1   No Provision for Alignment.....................   27
4.11.2   Leader Pins and Bushings   27
4.11.3   Taper Lock Between Each Cavity and
         Core   29
4.11.4   Taper Locks and Wedges   30
4.11.5   Taper Pins   31
4.11.6   Too Many Alignment Features   31

5 Before Starting to Design a Mold   33
   5.1   Information and Documentation   33
         5.1.1   Is the Product Design Ready?   33
         5.1.2   Are the Tolerances Shown?   33
         5.1.3   Are the Tolerances Reasonable?   34
         5.1.4   What are the Cycle Times?   34
         5.1.5   What is the Expected Production?   34
         5.1.6   What are the Machine Specifications?   35
                 5.1.6.1   Mechanical Features   35
                 5.1.6.2   Productivity Features.....................   37
                 5.1.6.3   Additional Requirements for Some Molds ...   39
   5.2   Start of Mold Design   42
         5.2.1   Mold Shoes   42
                 5.2.1.1   No Mold Shoe Used   42
                 5.2.1.2   Standard Mold Shoes   43
                 5.2.1.3   Home-Made Mold Shoes   44
                 5.2.1.4   Special Mold Shoes.......................   44
                 5.2.1.5   Universal Mold Shoes.....................   44
                 5.2.1.6   Mold Hardware   45
         5.2.2   Mold Drawings...................................   45
                 5.2.2.1   Assembly and Detail Drawings............   45
                 5.2.2.2   How Many Drawings and Views?   46
                 5.2.2.3   Arrangement of Views   46
                 5.2.2.4   Notes on Drawings   47
                 5.2.2.5   Additional Information on the Drawings.....   47
         5.2.3   The Stack Layout   48
                 5.2.3.1   Significant Cross Section   48
                 5.2.3.2   Will the Product Slide (Pull) out of the
                           Cavity?.................................   48
                 5.2.3.3   Will the Product Eject Easily from the
                           Core?   49

		5.2.3.4	Establishing the Parting Line.............	51
		5.2.3.5	Is the Cavity Balanced?	54
		5.2.3.6	Determining the Method of Cavity Construction	54
		5.2.3.7	Determining the Total Area of the Stack....	55
		5.2.3.8	Determining the Core Construction	56
	5.2.4	Selection of a Suitable Runner System		56
		5.2.4.1	Cold Runner, Single-Cavity Molds.........	56
		5.2.4.2	Cold Runner, 2-Plate Molds..............	57
		5.2.4.3	Cold Runner, 3-Plate Molds..............	58
		5.2.4.4	Hot Runner (HR) Molds..................	59
		5.2.4.5	Cold and Hot Runner Molds, in Combination	60
		5.2.4.6	Insulated Runner Molds	62
		5.2.4.7	Common Rules for Runner Systems	63
		5.2.4.8	L/t Ratio	65
	5.2.5	Venting		65
		5.2.5.1	What is a Vent?	65
		5.2.5.2	Design Rules that Apply to All Molds	66
	5.2.6	Ejection..		68
	5.2.7	Cooling..		69
		5.2.7.1	Purpose of Cooling a Mold	69
		5.2.7.2	Show Cooling Lines in Stack	71
		5.2.7.3	Screws	72
	5.2.8	Alignment of Stack		73
	5.2.9	Design Review		73
5.3	Preload...			73
5.4	Mold Materials Selection			74
	5.4.1	Effect of Expected Production		75
	5.4.2	Forces in Molds		75
	5.4.3	Characteristics of Steels and Other Mold Materials...		75
		5.4.3.1	Availability...........................	76
		5.4.3.2	Strength of Material....................	76
		5.4.3.3	Fatigue	76
		5.4.3.4	Wear	77
5.5	Stack Molds			78
5.6	Mold Layout and Assembly Drawings..........................			79
	5.6.1	Machine Platen Layout		79
	5.6.2	Symmetry of Layout, Balancing of Clamp		79
	5.6.3	The Views		80

5.6.4   Completing the Assembly Drawing   80
5.6.5   Bill of Materials (BoM) and "Ballooning"...........   81
5.6.6   Finishing Touches................................   81

6   Review and Follow-up   82

7   Typical Examples..   83
7.1   Containers or Other Cup-Shaped Products.................   83
7.2   Technical Products......................................   87
7.3   Mold with Fixed Cores...................................   88
7.4   Mold with Floating Cores   89
7.5   Molds with Side Cores or Splits.........................   90

8   Estimating Mold Costs.......................................   92
8.1   Need for Estimate   92
8.2   Precedents..   93
8.3   No Precedents...   93
8.4   Methods of Estimating   94
8.5   Mold Cost and Mold Price   95

9   Machining, Mold Materials, and Heat Treatment   97
9.1   Machining of Mold Components   97
9.2   Materials Selection 101
        9.2.1   Steel Properties................................ 102
9.3   Heat Treatment.. 106

Appendix 1 CAD/CAM (Computer-Assisted Design–Computer-
Assisted Manufacturing) 108

1 Introduction

I believe that a short history of injection molding will help in the understanding of what is required from a mold designer. After the Second World War, when plastics technology was beginning, there were no "mold designers." When a mold was needed, it was produced by artisans in tool and die maker shops, who were trying to expand into new fields. They were skilled in building accurate steel tools and dies, and the boss of such shops often worked closely with the molder, who understood better what was required. The molder sketched, often crudely, how the mold should look, and the boss, by closely supervising the machinists as they built the mold components, then by assembling and testing the molds himself (at the molder), built well-functioning molds. These were usually suitable for the, at that time, few existing plastics molding materials, and quite satisfactory for the (by today's standards, low) productivity expected from such molds. But over the years, many new and better plastics were developed, more suitable for the ever increasing variety of products, each often requiring different molding parameters. At the same time, the demand for increases in productivity became a high priority.

These increased demands of the traditional tool and die maker generated high specialization, and the "mold maker" was born. The mold maker was still essentially an expert in machining and assembling, and depended on the input from plastics materials suppliers on how to process these materials; also, the materials suppliers were not always knowledgeable enough, and depended on feedback from the molders regarding performance of the plastics they supplied. The molder was instrumental in the operating features the mold should have, and was often involved even in the selection of mold materials (steels, etc.). Eventually, all this information required to build a mold had to be shown on paper, both for the use of machinists in the shop and for assembling of the mold. The services of draftsmen or designers now became necessary, to relieve the boss from these time-consuming chores. Gradually, mold designers became the middlemen between the molder (the customer), the mold shop, and the plastics suppliers. The designers and sometimes the molders attended meetings and seminars to learn about new plastics and their expected processing requirements, and to apply their newly learned knowledge to the design of all molds.

Eventually, everything depended on the mold designer, who became solely responsible for the construction and functioning of the molds, and the mold maker reverted to just building the mold, per instructions given by the designer and as shown on drawings. At first, only assembly drawings were produced, with the more important dimensions shown, but gradually, in addition to complete assembly drawings, every mold part was detailed (except standard hardware items), complete with appropriate tolerances, so that any skilled machinist would be able to produce these components, and the boss returned to running the shop and was rarely involved in design problems. The molds could then be assembled by strictly following the assembly drawing, ideally, without need for adjustments ("fitting"). The mold was then ready for testing and production.

In earlier days, molds would be tested only at the molder, but, gradually, many mold makers acquired molding machines of various sizes for in-house testing, rather than shipping the molds to the molder, often interrupting his production if he had no suitable machine available at the time, and then shipping the mold back for adjustments if required. This shipping back and forth was costly and time-consuming; quite often, it had to be done not only once but several times. The investment in test machines proved not an expense but a saving for all parties involved, even though the cost of testing is added to the mold cost.

The mold designer must be involved in the testing of every mold, because this is where the most experience is needed, especially if the new mold does not function or perform as wanted, and revisions are necessary. It is important for the designer to insist that the molding technician not make any changes to the mold while it is being tested unless the designer is present; the only way future designs can benefit from these experiences is if the problems and solutions are properly recorded and the changes are documented on the drawings before they are made. A complete, comprehensive test report issued before the mold is shipped will greatly assist the molder when starting up the new mold.

This book provides the designer student, and perhaps even the advanced designer, with some ground rules for designing injection molds. It focuses on the "why," rather than going into the details of the design, the "how."

Quite often designers do things mechanically (especially with a CAD [computer-assisted design] program), following designs or methods used before, without questioning whether they are using the best approach to the problem. The mechanical approach can be useful and time saving as long as the precedent (the earlier example) is similar to the current job. But often, designers do not really understand why they copied what they did. It may have been the right thing for one plastic material, but not for another; it may have been suitable for a small production, but not for a large one; and so on.

Numerous new plastics have been developed over the last few years for specific applications, such as toys, housewares, packaging, electronics, electrical equipment, cameras, films, automotive, farming and aircraft components, furniture, clothing, and housing. Some of these plastics may require different production methods to arrive at the shapes required, such as compression and injection molding, blowing, extruding, thermoforming, and stamping. Some plastics can be shaped by more than one process, but in most cases, a mold is required to give the product the required form. Molds for low-pressures processing are easier to build than molds for high pressures, such as injection molds. (There is very little difference between injection molds for plastics and molds for die casting, i.e., the molding of liquid metals such as zinc.)

In the future, other plastics and other methods of processing and shaping them will be developed, but at the present time, injection molding seems to be the most common and economical method to produce plastic products, especially where large quantities are required.

1.1   Economics of Mold Design

Economics is often overlooked when this subject is taught. Every designer knows that the mold is a large expenditure and that its cost will affect the cost of the molded product. What designers often do not see is that this is only relative. Certainly, a simple mold, without all the "bells and whistles" will be less expensive, if the anticipated production run with the mold is relatively small. In some cases, it may be even of economic advantage not to mold a product completely as designed, but do some postmolding operations for those areas in the design that would require expensive features in the mold. For example, holes could be drilled after molding at an angle to the mold axis rather than designing and building complicated side cores; similarly, stamping of side wall could avoid a "split" mold. The designer must always consider the overall picture. It is more important to produce the lowest cost of the finished molded part, taking into account the cost of material, molding cost, and cost of direct labor involved in finishing the molded product, and including the cost of any postmolding equipment, such as drilling fixtures.

On the other hand, in real mass production, where many many millions of parts are expected to be produced, the mold should be built with the best mold materials and the best mold design features, always keeping in mind that the actual mold cost, even though higher, will have a negligible effect on the cost per unit. It should also be clear that there is a difference between mold making as part of the molder's operation and mold making as a business, that is, making molds for selling to a molder or end user. The molder may forgo some of the "appearance" features that would be expected from a reputable mold-making business. The molder will also be more aware of the expected production requirements and may take shortcuts that the mold maker in business would not. Today, most molders, but also many mold makers, specialize in certain areas. There are specialists for thin-wall molding, screw-cap making, large beverage container crates, preforms for PET bottles, small gears, and many others. This leads to the specialization of designers for the molds for these applications. But regardless of what size and type product is injection molded or who designs or builds the mold, the basic mold design principles as explained in this book are always the same. In this book, the designer should not look for pictures (drawings) of existing molds, but will learn instead the many things that must be considered when designing a mold. This does not mean that pictures of molds cannot be helpful, but every mold is different and some may require a better approach than the older mold depicted.

I will refer occasionally to three of my earlier books: *Understanding Injection Molding Technology (IMT), Mold Engineering (ME),* and *Understanding Product Design for Injection Molding (PD).*

2 Starting New in the Mold Design Field

The only prerequisite for the beginner is some knowledge of mechanical drawing delineation, whether it is done electronically on a computer (with programs like Autocad) or on the drawing board with pencil. Of course, the designer must also be familiar with some areas of basic arithmetic and trigonometry; both are required to put dimensions on the mold parts so they can be machined. Some of the advantages of electronic drafting are the following:

(1) Designs of entire, or portions, of earlier built molds can be easily used again by simply copying or modifying some existing design features from the program's memory, without the need for tedious redrawing.

(2) An up-to-date library of standard mold components and hardware can be established, which can be easily and quickly accessed and reproduced in new designs without the need for redrawing them every time they are needed.

(3) The quality of the drawings produced by a computer printout does not depend on the skilled hand of the designer.

(4) The computer permits easy transmission of designs to other locations, such as in-house manufacturing centers or manufacturers at other addresses.

Note the computer is only a tool to the designer; ultimately, the quality of a design depends entirely on how well the designer understands what is required and what can be made. Also be aware that even the most experienced designer will not always come up with the best design on the first attempt, but will try out different ideas in the course of the design job. This often necessitates sketching, erasing, and redrawing part or all of the picture, which is much easier to do electronically. There is a saying about the difference between a draftsman and a designer: "the draftsman uses the pencil, the designer uses the eraser." In the old days, the designer made his drawings on paper without much care for the appearance of the resulting picture; it was then usually left to draftpersons to produce a good, readable drawing.

The important thing is the thought that goes into the design of the mold, to ensure the best possible design. Different solutions are always possible to achieve the same end; in fact, all mold designers have their own ideas on how to solve certain design problems. To take advantage of various ideas, and to arrive at the best mold, it is good practice for the designer, after creating a mold layout, to consult with a colleague, or to arrange a design meeting of peers to discuss the proposed design. In many cases it is even better to provide two or more different layouts. These alternatives should then be discussed, and the best design or a composite of the various ideas should then be agreed upon.

This procedure is standard practice in all major design offices around the world. It may appear to be time-consuming, but the time (and emotions) invested in such peer critique are usually outweighed by the benefits of arriving at a better mold. Since, in general, mold designers (especially beginners) may not be familiar with machining and assembly practices, someone who is familiar in these areas should be included at such design meetings; this prevents a design of mold parts that may be difficult (or even impossible) to produce economically or to put together at assembly. It is also beneficial to have someone who knows the actual molding process look at a new layout. It is much less expensive to catch an error while it is still in the designing stage, than to find out about it later when steel has been cut or, even worse, when the mold is completed.

Time and money can be saved by spending more time during the design stage to consider alternatives and to get the designer involved in the manufacturing process of the mold, than by rushing a job through the design office to save a few hours there. When estimating the total time to build a mold, allocate approximately 15–20% of the total time for designing and detailing, about 60–70% for machining, and 15–20% for assembling the mold. (This, of course, depends on the shape of the product and the complexity of the mold.) And remember that the better the drawings are when given to the shop (or an outside source), the less time is wasted during machining and assembly of the mold.

3   The Basics of an Injection Molding Machine

(See also *IMT*, which contains much basic information on injection molding, molding machines, and molds.) The injection molding machine (Fig. 3.1) provides

- A safe support for the mold
- The opening and closing motion of the mold halves
- The clamping force to keep the mold closed while injecting
- The melted (plasticized) plastic to be injected
- The injection force to fill the mold cavity space
- The ejection force
- All necessary sequencing and temperature controls
- Any additional functions as may be required

Molding machines come in many different sizes, from small machines with a few kilonewtons (tons) of clamping force, to giant machines with 80,000 kN

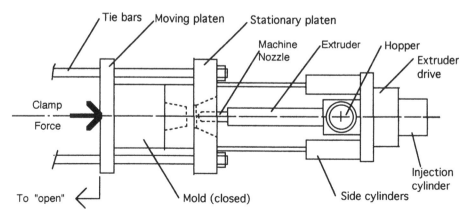

Figure 3.1   Schematic of an injection molding machine (top view).

(8800 US tons), for very large products. All machines can be equipped with a choice of standard injection units, suitable for the mold size and output required.

At this point, we will not go further into the functions of the molding machine. When discussing the injection mold, we will explain, when required, how the functions of the machine and the mold are interrelated.

4 Understanding the Basics of the Injection Mold

4.1 Design Rules

There are many rules for designing molds. These rules and standard practices are based on logic, past experience, convenience, and economy. For designing, mold making, and molding, it is usually of advantage to follow the rules. But occasionally, it may work out better if a rule is ignored and an alternative way is selected. In this text, the most common rules are noted, but the designer will learn only from experience which way to go. The designer must ever be open to new ideas and methods, to new molding and mold materials that may affect these rules.

4.2 The Basic Mold

4.2.1 Mold Cavity Space

The mold cavity space is a shape inside the mold, "excavated" (by machining the mold material) in such a manner that when the molding material (in our case, the plastic) is forced into this space it will take on the shape of the cavity space and, therefore, the desired product (Fig. 4.1). The principle of a mold is almost as old as human civilization. Molds have been used to make tools, weapons, bells, statues, and household articles, by pouring liquid metals (iron, bronze) into sand forms. Such molds, which are still used today in foundries, can be used only once because the mold is destroyed to release the product after it has solidified. Today, we are looking for *permanent* molds that can be used over and

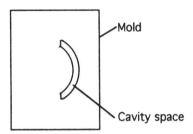

Figure 4.1   Illustration of basic mold, with one cavity space.

over. Now molds are made from strong, durable materials, such as steel, or from softer aluminum or metal alloys and even from certain plastics where a long mold life is not required because the planned production is small. In injection molding the (hot) plastic is injected into the cavity space with high pressure, so the mold must be strong enough to resist the injection pressure without deforming.

4.2.2   Number of Cavities

Many molds, particularly molds for larger products, are built for only 1 cavity space (a single-cavity mold), but many molds, especially large production molds, are built with 2 or more cavities (Fig. 4.2). The reason for this is purely economical. It takes only little more time to inject several cavities than to inject one. For example, a 4-cavity mold requires only (approximately) one-fourth of the machine time of a single-cavity mold. Conversely, the production increases in proportion to the number of cavities. A mold with more cavities is more expensive to build than a single-cavity mold, but (as in our example) not necessarily 4 times as much as a single-cavity mold. But it may also require a

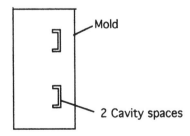

Figure 4.2   Illustration of basic mold with two cavity spaces.

larger machine with larger platen area and more clamping capacity, and because it will use (in this example) 4 times the amount of plastic, it may need a larger injection unit, so the machine hour cost will be higher than for a machine large enough for the smaller mold. Today, most multicavity molds are built with a preferred number of cavities: 2, 4, 6, 8, 12, 16, 24, 32, 48, 64, 96, 128. These numbers are selected because the cavities can be easily arranged in a rectangular pattern, which is easier for designing and dimensioning, for manufacturing, and for symmetry around the center of the machine, which is highly desirable to ensure equal clamping force for each cavity. A smaller number of cavities can also be laid out in a circular pattern, even with odd numbers of cavities, such as 3, 5, 7, 9. It is also possible to make cavity layouts for any number of cavities, provided such rules as symmetry of the projected areas around the machine centerline (as explained later) are observed.

4.2.3   Cavity Shape and Shrinkage

The shape of the cavity is essentially the "negative" of the shape of the desired product, with dimensional allowances added to allow for shrinking of the plastic. The fundamentals of shrinkage are discussed later.

The shape of the cavity is usually created with chip-removing machine tools, or with electric discharge machining (EDM), with chemical etching, or by any new method that may be available to remove metal or build it up, such as galvanic processes. It may also be created by casting (and then machining) certain metals (usually copper or zinc alloys) in plaster molds created from models of the product to be made, or by casting (and then machining) some suitable hard plastics (e.g., epoxy resins). The cavity shape can be either cut directly into the mold plates or formed by putting inserts into the plates.

4.3   Cavity and Core

By convention, the hollow (concave) portion of the cavity space is called the cavity. The matching, often raised (or convex) portion of the cavity space is called the core. Most plastic products are cup-shaped. This does not mean that they look like a cup, but they do have an inside and an outside. The outside of the product is formed by the cavity, the inside by the core. The alternative to the cup shape is the flat shape. In this case, there is no specific convex portion, and sometimes, the core looks like a mirror image of the cavity. Typical examples for this are plastic knives, game chips, or round disks such as records. While these items are simple in appearance, they often present serious molding problems for ejection of the product. Usually, the cavities are placed in the mold half that is mounted on the injection side, while the cores are placed in the moving half of the mold. The reason for this is that all injection molding machines provide an ejection mechanism on the moving platen and the products tend to shrink onto and cling to the core, from where they are then ejected. Most injection molding machines do not provide ejection mechanisms on the injection ("hot") side.

We have seen how the cavity spaces are inside the mold; now we consider the other basic elements of the mold.

4.4   The Parting Line

In illustrations Figs. 4.1 and 4.2 we showed the cavity space inside a mold. To be able to produce a mold (and to remove the molded pieces), we must have at least two separate mold halves, with the cavity in one side and the core in the other. The separation between these plates is called the parting line, and designated P/L. Actually, this is a parting area or plane, but, by convention, in this context it is referred to as a line. In a side view or cross section through the mold, this area is actually seen as a line (Fig. 4.3).

The parting line can have any shape, but for ease of mold manufacturing, it is preferable to have it in one plane. The parting line is always at the widest circumference of the product, to make ejection of the product from the mold possible. With some shapes it may be necessary to offset the P/L, or to have it at

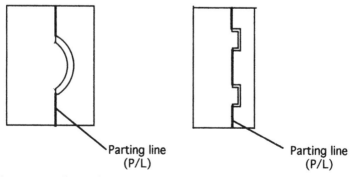

Figure 4.3   Illustration of schematic mold, showing the parting line.

an angle, but in any event it is best to have is so that it can be easily machined, and often ground, to ensure that it shuts off tightly when the mold is clamped during injection. If the parting line is poorly finished the plastic will escape, which shows up on the product as an unsightly sharp projection, or "flash," which must then be removed; otherwise, the product could be unusable. There is even a danger that the plastic could squirt out of the mold and do personal damage.

4.4.1   Split Molds and Side Cores

There are other parting (or split) lines than those that separate the cavity and core halves. These are the separating lines between two or more cavity sections if the cavity must separate (split or retract) to make it possible to eject the molded product as the mold opens for ejection.

Figure 4.4 shows simple "up and down" molds. The machine clamping force holds the mold closed at the P/L. (In (B) and (C), the parting line could be anywhere on the outside of the rim, between the two positions shown, but is preferred as in (B).) In (D) we must consider the injection pressure p (as shown with small arrows inside the cavity space), which will force the two cavity halves in the direction of the the large arrow m. This force also exists in the other examples, but is resisted by the strength of the solid cavity walls, which do slightly expand during injection and then return to their original shape once the injection cycle is completed. Since these side forces can be considerable (see Section 4.6), the mold plates (the "mold shoe") must be sufficiently solid to

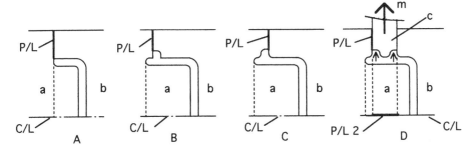

Figure 4.4   Schematic illustrations of location of parting lines (P/L) (only one half of mold shown): (a) core, (b) cavity. (A) Simplest case: P/L at right angles to axis of mold. (B and C) Product with rim but still simple. P/L can be either as in (B) or in (C). (D) Simple product but with rim and projection. Cavity is split, creating an additional P/L 2.

contain these forces and provide the necessary preload to prevent opening of the mold during injection. These side cores, or split portions of the cavities, can represent just small parts of the cavity, or even only small pins to create holes in the side of the products, but they could also be sections molding whole sides of a product, as, for example, with beverage crates or large pails.

4.5   Runners and Gates

In Fig. 4.3, we showed molds with cavity spaces and parting lines. Now, we must add provisions for bringing the plastic into these cavity spaces. This must be done with enough pressure so that the cavity spaces are filled completely before the plastic "freezes," that is, cools so much that the plastic cannot flow anymore. The flow passages are the sprue, from where the machine nozzle (see Fig. 3.1) contacts the mold, the runners, which distribute the plastic to the individual cavities, and the gates, which are (usually) small openings leading from the runner into the cavity space. We discuss the great variety of sprues, runners, and gates later. We illustrate here only two methods of so-called cold runners (see Fig. 4.5).

The left part of Fig. 4.5 shows the simplest case of a single-cavity mold, with the plastic injected directly from the sprue into the cavity space. This is a frequently used method, mostly with large products. It is inexpensive, but requires the clipping or machining of the relatively large (sprue) gate. The right drawing is of a typical (2-plate) cold runner system, with the plastic flowing through the sprue and the runner and entering the cavity space through relatively small gates, which break off easily after ejection. Instead of the 2 cavities as shown here, there can be any number of cavities supplied by the cold runners. These and other runner methods are explained later.

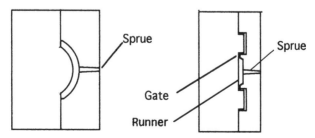

Figure 4.5   Illustration of schematic mold, showing cold sprue (left) and cold runner (right).

4.6   Projected Area and Injection Pressure

At this point we digress and consider injection pressure and how it affects mold design (see Fig. 4.6). As the plastic fills the cavity space under high pressure p, the pressure, in the direction of the mold (and machine) axis—in other words, in the direction of the motion of the clamp—will tend to open the cavity at the parting line. The separating force F created by the pressure p is equal to the product of the pressure p times the projected area A, which is the area of the largest projection of the product at the parting line. The arrow describing projected area in Fig. 4.6 really describes an area not a line, as delineated in this section view of the mold. The actual area can be seen (and measured) in a plan view of the mold cavity. From this it becomes clear that the clamping force, the force exerted on the mold by the molding machine, must be at least as great as the force F to keep the mold from opening (cracking open) during injection.

The difficulty is how to determine the value of the injection pressure p. We can easily calculate the injection pressure inside the machine nozzle, which is directly related to the size of the injection cylinder of the machine and the hydraulic (oil) pressure supplying the injection cylinder. The injection pressure at the machine nozzle, in general, is adjustable between any low values, to a high of about 140 MPa (20,000 psi), in most molding machines, and in some machines can be as high as 200 MPa (29,000 psi) or even higher. This pressure,

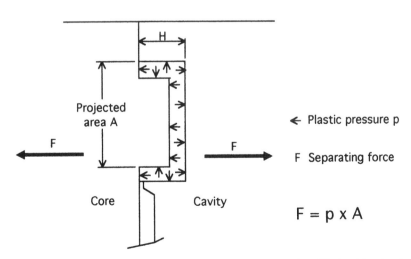

Figure 4.6   Portion of a schematic mold, showing a cavity filled with plastic under pressure acting in all directions.

however, is greatly reduced (by the pressure drop) by the time the plastic passes through the machine nozzle orifice, the runners, and the gates, and as it flows through the narrow passages of the cavity space. The flow also depends largely on the viscosity (defining the ease of flow) of the plastic, which depends on its chemistry and on its temperature (the higher the temperature, the lower the viscosity). This area is the subject of much research and experimentation, and computer programs are available to calculate the pressures and the flow inside the cavity space (see Appendix).

A good working assumption is a cavity pressure p of approximately 30–40 MPa (4000–5000 psi) for average product wall thicknesses of about 2–3 mm or more, and 40–50 MPa (5000–6000 psi) or even higher for thin-wall products. For example, a disk of 100 mm (10 cm) diameter, with a thickness of 2 mm, will generate an opening force of $(10^2 \times \pi \div 4)\,\text{cm}^2 \times 30\,\text{MPa} = 235\,\text{kN}$ (approx. 26 US tons) per cavity.

4.6.1   Clamping Force

From the above example we see that a clamping force of at least 235 kN (26 US tons) per cavity should be used to ensure that the mold will not crack open. If the average wall of the product is thinner, or if the definition, that is, the accuracy and clarity of reproduction of details in the cavity wall, is important, then the pressure must be higher and a larger clamping force will be required.

4.6.2   Strength of the Mold

There are two other serious effects of the injection pressure p. First, as can be seen in Fig. 4.6, the pressure also acts in the direction at right angles to the axis of the mold. These forces, which are the product of the projection of the cavity in this direction times the pressure p, will tend to stretch and deflect the cavity walls outward. The greater the height H of the product, the greater will be this force and the stronger must be the walls surrounding the cavity.

Second, the clamping force is applied as soon as the mold closes. At this moment, the whole clamp force is resisted ("taken up") by the area of the land, which is the area surrounding the cavity that touches the core side. If this area is too small, the land will be crushed and damage the sealing-off surfaces of the parting line, eventually ruining the mold. Proper sizing of the land and correct materials and hardness (steel, etc.), or other measures to counteract the clamping forces are the solution to this problem. Also, the mold setup technician should be informed by a nameplate attached to the mold that the recommended maximum clamp force for the mold must not be exceeded during mold setup or during operation.

4.6.3　Why Are High Injection Pressures Needed?

High injection pressures are needed to ensure that the mold is completely filled during the injection cycle, with the desired clear surface definition. There are several problems to consider.

(1) The thinner the wall thickness of the product, the more difficult it is to push the plastic through the gap between cavity and core, thus requiring higher pressures. Since material (the plastic) usually accounts for 50–80% of the total cost of a molded product, it is highly desirable to reduce the weight (mass) of plastic injected to a bare minimum. This usually means reducing the wall thickness as far as possible without affecting the usefulness of the product. Over the years, many products have been redesigned just to reduce the plastic mass of a product. This is also why many modern injection molding machines provide higher injection pressures than older ones.

(2) The colder the injected plastic, the higher its viscosity, and the more difficult it becomes to fill the mold. The cost of the product depends directly on the cycle time required to mold a product. The higher the melt temperature of the plastic, the easier it will flow and fill the mold. However, higher melt temperatures also require increasing the cooling cycle time to bring the temperature of the injected plastic down to a level where the product can be safely ejected without distorting or otherwise damaging it. This means more power (for heating and cooling), longer cycles, and therefore higher costs. It is often better to inject at the lowest possible temperatures, even if more pressure is needed to fill the mold. Note that higher injection pressures will require greater clamping forces and a stronger, possibly larger, machine. Another solution to the problem might be to select a plastic that flows more easily. Such plastics, however, are usually more expensive and may not be as strong as desired.

(3) High injection forces are needed for good surface definition. Typically, this is important when molding articles such as compact discs, where the clarity

and precision of the surface definition is in direct relation to the quality of the sound reproduction of the recording.

4.7   Venting

As the plastic flows from the gate into the cavity space, the air trapped in it as the mold closed must be permitted to escape. Typically, the trapped air is being pushed ahead by the rapidly advancing plastic front, toward all points farthest away from the gate. The faster the plastic enters—which is usually desirable— the more the trapped air is compressed if it is not permitted to escape, or vented. This rapidly compressed air heats up to such an extent that the plastic in contact with the air will overheat and possibly be burnt. Even if the air is not hot enough to burn the plastic, it may prevent the filling of any small corners where air is trapped and cause incomplete filling of the cavity. Most cavity spaces can be vented successfully at the parting line, but often additional vents, especially in deep recesses or in ribs, are necessary.

Another venting problem arises when plastic fronts flowing from two or more directions collide and trap air between them. Unless vents are placed there the plastic will not "knit" and may even leave a hole in the wall of the product. This can be the case when more than one gate feeds one cavity space, or when the plastic flow splits in two after leaving the gate, due to the shape of the product or the location of the gate. Within the cavity space, plastic always flows along the path of least resistance, and if there are thinner areas, they will fill only after the thicker sections are full.

Venting is discussed more thoroughly in *ME*, Chapter 11.

4.8   Cooling

Cooling and productivity are closely tied. In injection molding, the plastic is heated in the molding machine to its processing (melt) temperature by adding energy in the form of heat, which is mostly generated by the rotation (work) of the extruder screw. After injection, the plastic must be cooled; in other words, the heat energy in the plastic must be removed by cooling, so that the molded piece becomes rigid enough for ejection. Cooling may proceed slowly, by just letting the heat dissipate into the mold and from there into the environment. This is not suitable for large production, but for very short runs "artificial" cooling of a mold is not always required. However, for a production mold, good cooling to remove the heat efficiently is very important.

4.8.1   Basics of Cooling

The physics and mathematics of cooling are quite complicated. Computer programs can determine the appropriate means of cooling a particular mold, after input of the geometry of the product and the mold, and based on assumed temperatures of melt and coolant, flow patterns and sizes of the cooling channels, and other variables, such as heat characteristics of the coolant and the mold materials. This means that a computer program can determine the best planned cooling layout for a mold only *after* the mold is designed. But the designer wants to know how to design the best cooling layout in the first place. There are several rules, based on experience, to help the designer.

- **Rule 1:** Only moving coolant is effective for removing heat. Stagnant coolant in ends of channels, or in any pocket, does nothing for cooling.

- **Rule 2:** All cavities (and cores) must be cooled with the same coolant flow (quantity of coolant per unit of time) at a temperature that is little different from cavity to cavity (or core to core). The coolant temperature will rise as it passes through each cavity (or core), but this is the very purpose of the coolant: to remove heat, which will raise its own temperature. As long as the temperature difference ΔT between the first and the last cavity in one group of cavities (or cores) is not too large—on the order of $\Delta T = 1-5\,^{\circ}\text{C}$ (2–9 °F), depending on the job—the system is working properly. The smaller the difference, the more coolant will be required (which is more expensive in operation). In many molds there can be a good argument for compromise by having a greater ΔT and thereby using less coolant. In some cases, however, the lowest ΔT value may be necessary for quality requirements of the product. This may require special coolant capacity and pumps.

- **Rule 3:** The amount of heat removed depends on the quantity (volume) of coolant flowing through the channels in cavity (or core). The faster the coolant flows, the better it is, because (a) a greater volume will flow through the channels, and (b) there will be less temperature rise of the coolant from the first to the last cavity (or core).

- **Rule 4:** The coolant must flow in a turbulent flow pattern, rather than in laminar flow. Turbulence within the flow causes the coolant to swirl around as it flows, thereby continuously bringing fresh, cool liquid in contact with the hot metal walls of the cooling channels, and removing more heat. By contrast, laminar flow moves along the channel walls relatively undisturbed, so that the outer layer of the coolant in touch with the metal will heat up, but the center of the coolant flow will remain cold, thus doing little cooling.

Turbulent flow is defined by the Reynolds number (Re), which is calculated as $Re = (V \times D) \div v$, where V is the velocity of the coolant (m/s), D is the diameter of the channel (m), and v is the kinematic viscosity (m^2/s). $v = \mu \div \rho$, where μ is the absolute viscosity (kg/m · s), and ρ is the density of the coolant (kg/m^3). A Reynolds number of more than 4000 (Re > 4000) designates turbulent flow. The higher the number, the better the cooling efficiency. For good cooling, $10,000 < Re < 20,000$ should be attempted. For water at 5 °C (41°F), $\rho = 999.5$ kg/m^3, $\mu = 1.55 \times 10^3$ kg/m · s, and $v = 1.5508 \times 10^{-6}$ m^2/s. (More values can be found in *ME*, in Table 25.2.)

Thus, where cooling is important—in cavities, cores, inserts, side cores, and so on—small-diameter channels and fast-flowing coolant are also important. Most cooling lines for cavities and cores are supplied from channels in the underlying or surrounding plates, and can be much larger, therefore having a much smaller Re number. But this is usually satisfactory because these plates do not need as much cooling as the stack parts, which come in contact with the hot plastic.

■   **Rule 5:**   Serial or parallel flow? (See Fig. 4.7.) It does not matter whether the coolant follows a serial flow, that is, from cavity to cavity (or core to core) in sequence (Fig. 4.7a), or whether the flow is split so that the coolant flows in a parallel pattern (Fig. 4.7b), as long as each branch has the same flow. In many multicavity molds, the cooling channels are arranged so that they are partly in parallel and partly in series (Fig. 4.7c). Often, in the same mold, cavities are in one arrangement of series, parallel, or both, and cores, inserts, or side cores, are in another arrangement, whichever is more suitable for the layout. There is no rule for which way to go, as long as the flow rules are followed.

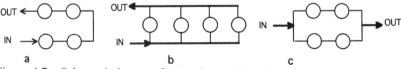

Figure 4.7   Schematic layout of (a) series cooling, (b) parallel cooling, and (c) series–parallel cooling.

■   **Rule 6:**   The channel sizes (cross sections) must be calculated so that there is always more than enough flow capacity in a preceding section to feed equally all the channels in the following split, parallel sections. For example, if there are 4 parallel channels of $40\,\text{mm}^2$ cross-sectional area each, the (preceding) feeder must have at least $4 \times 40\,\text{mm}^2 = 160\,\text{mm}^2$ cross-sectional area. In some molds there are 4 or more points where the cross sections step down in the cooling system. It does not matter if the preceding section is greater than the calculated minimum value, but it must not be smaller, if the coolant is to flow equally through all subsequent channels. Coolant, like plastics, always takes the path of least resistance. For example, if the preceding cross section is $3x$, and each of 4 succeeding parallel cross sections are x, there will not be enough coolant, and one of the 4 channels will see little or no flow through it. Unfortunately, this is often missed in designs and the mold does not function properly.

■ **Rule 7:** The difficult-to-cool areas in the mold must be considered first. These are, essentially, all delicate mold features, such as thin and slender core pins, blades, and sleeves. Slender signifies, in this context, that the ratio of length over the narrow bottom dimension or diameter of a pin or insert is more than 2 to 1. Remember that heat always flows from the higher toward the lower temperature; the flow decreases as the length of travel increases and as the cross-sectional area through which the heat travels gets smaller. Difficult-to-cool areas limit the mold cooling capability and seriously affect the molding cycle. There is no sense in providing good cooling for the easy-to-cool areas of the mold if there are poorly cooled areas elsewhere in it. Selecting materials such as beryllium–copper alloys may help to remove the heat faster, or special cooling methods may be used, such as blowing (cold) air at the thin sections while the mold is open. But first the designer must try to find a way of getting coolant (not necessarily water) into the thin sections, or at least get the best cooling into the mold parts supporting these thin projections.

■ **Rule 8:** Study the product to locate heavy sections of the plastic. They are always a problem, even where it is easy to provide good cooling, because of potential shrink and sink marks. Heavy sections are particularly bad if they are toward the end of the plastics flow where there is less pressure to ensure good filling. The mold designer should discuss this problem with the product designer. There may be the possibility of a minor alteration of the product design to avoid heavy sections so that not only is plastic saved but also cooling time is reduced. For example, the heavy, solid handle of a coffee mug could be redesigned by coring it from both sides. This could add to the mold cost, but would greatly reduce the cycle time. The question is whether the customer wants to sacrifice design features for productivity. (See also *Understanding Product Design for Injection Molding*.)

4.8.2   Plate Cooling

An often overlooked fact is that mold cooling is not only for cooling the plastic, but also for cooling the various mold plates that are close to areas heated by the plastic, such as the hot runner systems discussed later or, in special cases, such as injection blow molding, where the mold cores are heated to keep the plastic hot, for blowing immediately after injection. As is explained in Section 4.10, all materials expand when heated. In many molds, certain plates are essential for the alignment system because they carry the leader pins and bushings or other alignment members. If the mold plates are at different temperatures, they will expand differently from their original, cold state, and cause misalignment between the alignment elements. For example, assume that the distance of two leader pins in a mold is $L = 400$ mm and that a temperature difference of $\Delta T = 10\,°C$ (18 °F) exists between the two plates carrying the pins and bushings. With an approximate heat expansion for steel of 0.000011 mm/mm/°C, L will increase by ΔL. $\Delta L = L \times \Delta T \times 0.000011 = 400 \times 10 \times 0.000011 = 0.044$ mm (0.00173 inch). Considering that the standard diametrical clearance between leader pins and bushings is only 0.025 mm (0.001 inch), the example shows the pins will bend at every cycle, or bind in the bushings. This points to the importance of ensuring in the design that both mold halves should be kept as close as possible to the same temperature. (Compression molding, usually employed for thermosetting materials, requires heating of the mold, regardless of productivity. In this process, the plastic must be heated to set (or harden); the product leaves the mold hotter than the raw material used to fill the mold.)

More about cooling later. See also *ME*, Chapter 13.

4.9   Ejection

After the plastic in the cavity spaces has cooled sufficiently and is rigid enough and ready for removal, the mold halves move apart, allowing sufficient space between the mold halves for removal of the product. As with cooling, the complexity of any provision for ejection from the mold is a question of the desired productivity. Some products don't need any provision within the mold for ejection. For example, a quick blast from an air jet applied manually by an operator and directed at the parting line can lift a (simple) product off the core or out of the cavity, but this would not be practical in most molds, and is rarely used for real production. Usually, the products are ejected by one of the following methods:

(1) Pin (and sleeve)
(2) Stripper plate or stripper ring
(3) Air alone
(4) Air assist
(5) Combination of any of the above (1), (2), (3), and (4)
(6) Unscrewing, in case of screw caps, etc.
(7) Combination of any of the above, combined with robots

The most common and oldest methods are

- Pin (and sleeve) as shown in Fig. 4.8
- Stripper plate or stripper ring, as shown in Fig. 4.9

  These two systems can be used in most molds and for most plastics. The problem with both these systems is that there are heavy moving parts involved, and the upkeep of such molds is high.

- Air ejection alone can be used for flat products (Fig. 4.10, *left*), but for deep cup-shaped products (*right*) it is restricted to only certain plastics and shapes. The main advantage is that it has no, or almost no, moving

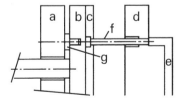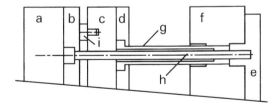

Figure 4.8   (*Left*) Section through ejector pin mold: (a) backing plate, (b) ejector plate, (c) ejector retainer plate, (d) core plate, (e) molded product, (f) ejector pin, (g) stop pin. (*Right*) Section through sleeve ejector mold: (a) backing plate, (b) core pin retainer plate, (c) ejector plate, (d) sleeve retainer plate, (e) molded product, (f) core plate, (g) sleeve ejector, (h) core pin, (i) stop pin.

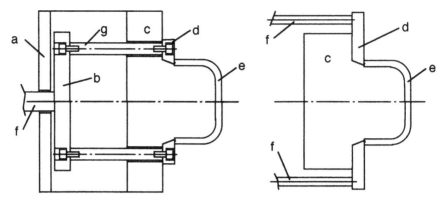

Figure 4.9 (*Left*) Section through stripper ring mold: (a) mounting plate, (b) ejector plate, (c) core plate, (d) stripper ring, (e) molded product, (f) machine ejector, (g) connecting sleeve. (*Right*) Section through stripper plate mold: (c) core and mounting plate, (d) stripper plate, (e) molded product, (f) machine ejectors.

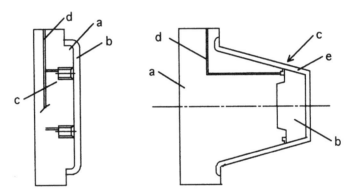

Figure 4.10   Air ejection alone. (*Left*) (a) core and mounting plate, (b) molded product, (c) air valves, (d) pressure air supply. (*Right*) (a) core and mounting plate, (b) core tip, (c) circular air gap, (d) pressure air supply, (e) molded product.

parts. Air ejection alone is often used in very high production molds; the same applies to (7), by combining any of the above ejection methods with integrated robots.

Note that for best productivity, to reduce cycle time, the products should be ejected as early as possible. Certain ejection methods permit earlier ejection; others depend on the plastic to be stiffer. For example, stripping permits hotter (softer) products to be ejected without damage to them, whereas unscrewing requires the pieces to be more rigid.

4.9.1   Automatic Molding

Earlier molds were all designed to require operators (often lowly paid and unskilled) to sit or stand at the molding machine. After every cycle they opened the safety gate to remove the products from the molding area, reclosed the gate and initiated the next molding cycle. They also were, in some cases, supposed to visually inspect the products at this time and even make adjustments to the machine if they thought it necessary. Because the molds were often not properly finished, by today's standards, or had unreliable injection and ejection systems, the operator was also often required to reach into the molding area to pry loose a stuck, possibly defective product, and from time to time had to lubricate the molding surfaces with mold release agents. All this was not only labor intensive, adding greatly to the cost of production, but was also very unsafe and the cause of many serious injuries. Since much of this operation also depended on the acquired skill of the operator—some workers are faster, some slower—and on the time of the day or night, or even on the day of the week, the overall molding cycle time could vary considerably, resulting in quality differences of the product because of different residence times of the melt in the machine; many rejects resulted. There was also the problem of absenteeism of the personnel, which often played havoc with production planning. Much effort was therefore spent on eliminating operators from the actual molding process.

Fully automatic (FA) molding depends essentially on two factors:

(1) *Reliable injection.* The molding machine must be repetitive from cycle to cycle in every aspect, but especially in the dosing (the amount of plastic injected) and the melt temperature.

(2) *Reliable ejection.* This is 100% the responsibility of the mold designer. Every mold (with very rare exceptions) can be designed so that there is no chance of the product hanging up and not ejecting. The key to good ejection is that the product always stays on the side from which it will be ejected, usually, but not necessarily, from the core side of the mold. The designer must select the appropriate method of ejection and make sure that there is enough ejection stroke to clear the products from the cores. This is frequently overlooked and can also be caused by improper setup of the mold. Many areas must be considered in the design; some are discussed later.

The designer must keep in mind Murphy's law, which says that if it can happen, it will.

See also *ME*, Chapter 12.

4.10   Shrinkage

One of the most misunderstood areas of mold design is shrinkage. Every material (metals, plastics, gases, liquids) expands as its temperature increases (heat expansion) and returns to its original volume if cooled down to the original temperature. The problem with all plastics is the characteristic of compressibility. All solid materials compress under load, but most not as much as plastics. When pressure is applied to plastics (or to hydraulic oil, but not to water), plastics will compress significantly (i.e., reduce in volume) in proportion to the amount of pressure applied. This may be (within the range of molding operations) as high as 2% of the original volume. Thus, we now have two conditions that work against each other: heat expansion and compressibility. As the plastic is injected, it is both hot and therefore expanded, but also under significant pressure, which reduces its volume. This makes it very difficult to arrive at a true shrinkage factor, because the actual change in volume depends on the type of plastic, the melt temperature, the injection pressure required to fill the cavity space, and the temperature at which it will be ejected from the mold.

For practical purposes, and for many products and molds, the shrinkage factors supplied by materials suppliers can be used. However, these figures indicate only a range within which to choose, usually between 0 and 5%. In some cases, where the volume or size of a product is important, this is not accurate enough. With crystalline plastics, such as polyethylene (PE), polypropylene (PP), and polyamide (nylon), the shrinkage factor is much higher than with amorphous plastics, such as polystyrene (PS) and polycarbonate (PC). Plastics filled with inert substances, such as glass or carbon fibers or talcum, have a much lower shrinkage than that for the same but unfilled material. Shrinkage figures should be obtained from materials suppliers, for guiding purposes.

4.10.1   Variable Shrinkage

The designer must understand that the areas within the cavity spaces close to the gate see higher pressures, so the shrinkage there will be less and will require a smaller shrinkage factor. Conversely, near the end of the flow through the narrow cavity space, the pressure in the plastic is much lower than near the gate, and a higher shrinkage factor will apply. In some applications, more than two shrinkage factors may have to be selected within one cavity. It is also important to establish at what temperature the product will be ejected. If it is ejected while still hot, it will shrink more outside of the cavity space as it cools to room temperature. If ejected later, when it is cooler, it will shrink less, as measured in comparison with the steel sizes of the cavity and core.

This is sometimes, but uneconomically, used to arrive at the proper size of a product such as a container or lid. If a molded product is too small because not enough shrinkage value was added to the product dimensions when specifying the mold steel dimensions, the proper product size can be achieved by ejecting it later, when it is cooler, but this means loss in productivity. With high production, the proper procedure is to resize the steel dimensions.

See also *ME*, Chapter 8.

4.11   Alignment

Various methods are used to align cavity and core plates. The method selected depends on the shape of the product, the accuracy (or tightness of tolerances) of the product, and even on the expected mold life. Several choices are available:

(1)   No provision for alignment within the mold
(2)   Leader pins and bushings
(3)   Taper lock between each cavity and core
(4)   Taper lock between a group of cavities and cores
(5)   Wedge locks
(6)   Taper pins
(7)   Combination of (2) with (3), (4), (5), or (6)

4.11.1   No Provision for Alignment

In the case of a flat product, without any cavity (depression) in one mold half, and the cavity entirely in the other mold half, for example, in a mold for a floor mat, there is no need for alignment, even if there is some engraving on the flat surface of the mold, because the most the dimensions can vary is by the amount of play between the machine tie bars and the tie bar bushings.

4.11.2   Leader Pins and Bushings

This common method of alignment between mold halves is shown in Fig. 4.11. In cup-shaped products with heavy walls, there is really no need for alignment within the mold, because the clearances between tie bars and their bushings are usually much less than the tolerances of the product wall thickness. The main reason to have leader pins in these cases is to protect the projecting cores from physical damage, when handling the mold.

The protection of the cores by use of leader pins applies also to all other mold alignment methods. Wherever leader pins are used, they should be placed at the same mold side as the cores and be longer than the longest projection of the cores to protect them from damage (see dimension s, in Fig. 4.11). There are exceptions to this rule, for example, in some 3-plate molds.

What is often missed is that for most applications leader pins and bushings are a very accurate method of alignment. Consider dimension t in Fig. 4.11, and let's assume a wall thickness $t = 1.50$ mm (0.060 inch), with a tolerance of ± 0.05 mm (0.002 inch), or 1.50 ± 0.05 mm. With standard commercial hardware, the leader pin is usually nominal size minus 0.025 mm (-0.001 inch), and the bushing is nominal size plus 0.025 mm ($+0.001$ inch). Therefore, with one set of pins and bushings, the maximum clearance, in the highly unlikely worst case, between one set of leader pins and bushings could be 0.05 mm (0.002 inch) on the diameter, so the centers would be misaligned only half that amount. By having at least 2, but usually 4 sets, the total clearance between the pins in all the bushings would be even less. In the worst case, the

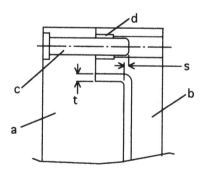

Figure 4.11   Typical mold with leader pin and bushing alignment: (a) core plate, (b) cavity plate, (c) leader pin, (d) leader pin bushing, (s) safety distance of pin above core, (t) wall thickness of plastic product at parting line.

possible play and misalignment would be well within the tolerance limits specified in this example, and therefore acceptable.

It can be easily seen that this holds true as long as the product has not much smaller wall thicknesses, as is often the case with thin-wall containers, with wall thicknesses in the order of 0.4 mm (0.015 inch) or even less. In those special but frequent cases, other methods of alignment must be used such as taper fits. We also must not forget the influence of heat expansion of the mold plates, which will affect the alignment accuracy.

4.11.3   Taper Lock Between Each Cavity and Core

Figure 4.12 shows 3 possible configurations of taper or wedge locks. On the left, the tapers in both male and female members match perfectly. Because of manufacturing tolerances, this is impossible to achieve except, perhaps, by individual fitting of parts, and even then it is difficult. To be able to produce any mold part without need for fitting (center), they must be closely toleranced and accurately machined. To solve the problem of providing proper alignment, the matching parts are dimensioned such that the male member is slightly larger than the female member, and the female member will be slightly expanded from the moment the mold halves touch, until the mold is fully clamped. The amount that the pieces stay apart before final clamping (d) is called preload in Fig. 4.12. This amount *d* is very, very small, and depends on the length of the taper and on its angle. It must be greater than zero. On the right, the female member is larger than the male member. This taper lock is useless because the tapers don't touch (f); no force is generated to pull the mold halves into alignment.

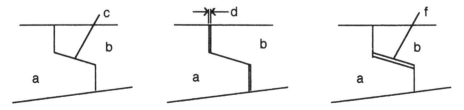

Figure 4.12   Taper (or wedge) lock: (a) male member, (b) female member, (c) taper. (*Left*) Ideal condition. (*Center*) Correct application. *d* is called preload. (*Right*) Useless taper.

In practice, it can be easily seen on a mold if the tapers work: If the tapers (or wedges) are shiny all around, they work; if they are rusty, or just dirty, they don't work, and the mold probably depends on the tie bars and tie bar bushings for alignment, or on the mold leader pins and bushings. It is surprising how many molds are in this category. Many times the designer (or the mold maker) thought that by providing tapers, the mold will be more accurately aligned. In most of these cases, the taper fit was wasted money. Note that working tapers are subject to severe wear and must be made from suitable, hardened steels, and even so will have to be replaced or repaired from time to time. Any size taper is acceptable, between 5 and 20°. (Common tapers are 7, 10, and 15°.) Too small a taper may cause locking and separation difficulty because of friction in the tapers; too large a taper requires too much force to close. Obviously, to move the tapers for the preload distance d, until they seat properly, means that the matching, female taper will have to be spread. This requires considerable force. When considering the clamp force of the machine, this must be considered and the forces calculated, especially with multicavity molds in which every stack is aligned with taper locks. If too much force is required for closing the mold, there may not be enough clamp force left for holding the mold closed during injection.

4.11.4   Taper Locks and Wedges

Taper locks are conical (usually round) matching mold parts, and the taper of the cone is designed to provide the alignment between two mold parts (cavity–core, core–stripper ring, etc.). This method is very accurate and relatively inexpensive, but has two inherent disadvantages:

(1)   The alignment of the various components depends on the accuracy of machining and once the assembly is finished, there is no possibility of adjusting the alignment.

(2)   Once the tapers wear, which is unavoidable due to the very nature of this design, which must touch and rub, they are difficult to repair and reuse without changing other mold parts as well. The easiest way is often to replace the worn elements.

Wedges are pairs of hardened, flat bars, with one side tapered. Four sets of wedges are always required per alignment, either for each cavity, or for the whole mold. The advantage is that wedges can be shimmed or ground on opposite pairs to adjust for wear or for inaccurate manufacturing, or easily replaced if shimming is not practical. The disadvantage of wedges is that they require more space on the mold surface, so the mold size will be larger than when using taper locks.

4.11.5   Taper Pins

Taper pins (and bushings) are sometimes used for the final alignment of cavity and core in addition to leader pins, where it is believed that the accuracy of leader pins is insufficient. They act similarly to taper locks and are available as standard mold hardware. It is questionable whether they do any better job than the other methods of alignments explained here; and they are subject to the same problems as taper locks, regarding wear and accuracy of machining the mold and/or core plates.

4.11.6   Too Many Alignment Features

Another problem is frequently encountered in poorly designed molds. Typically, cavities and cores can be aligned by either leader pins and bushings, or taper (or wedge) locks. Where high accuracy in alignment is required, taper (or wedge) locks are the preferred choice. However, they do not assure that the mold halves will stay together when handling the mold; there is always the danger that the cores and cavities could be damaged if the mold halves should separate and bang together once the taper engagement is lost. It is therefore necessary to equip the mold with leader pins (but not necessarily with leader pin bushings), in addition to the taper locks. Since the tapers will determine the final alignment, the leader pins must fit only loosely in their corresponding openings (or leader pin bushings) without actually contributing to the final alignment of cavities and cores. Quite often, even for large molds, only two such pins need to be provided, usually located at the top of the mold on the core side.

Similarly, some multicavity molds are built with small leader pins (usually only two) and bushings for each set of cavity and core and are mounted on the stack plates; they ensure the final alignment of each stack. In addition, two or four large leader pins are used to align the complete mold halves, but these pins also must be "loose" in their bushings, to prevent "fighting" between the two separate sets of alignments. An exception to this rule of loose pins is when a more expensive but superior method is used: the cores are mounted such that they can move slightly (float) on their backing plates; as the mold closes, the final alignment (tapers or pins) will move each core into position relative to its cavity. In this case, the leader pins mounted in the mold shoe (on the core side) will have their regular, standard clearances.

5   Before Starting to Design a Mold

5.1   Information and Documentation

Before starting to design a mold, the designer must make sure that all the information is on hand.

5.1.1   Is the Product Design Ready?

It is frustrating and wastes valuable time to find during your work that information is missing, or when significant changes are made after starting that can affect the concept of the mold.

5.1.2   Are the Tolerances Shown?

Are the dimensional tolerances specified on the drawing the same as when the mold cost was first estimated and the mold price quoted? This can have serious implications, especially if no tolerances were shown when the job was quoted; for example, if a molder requests an approximate mold cost so that he can estimate the final cost of the product for his customer. Unfortunately, sometimes there is not even a drawing, just a sample or model of the product used for the estimate.

While it is desirable that the mold designer is involved in the product design, to ensure that the product can be easily molded and will be satisfactory for the purpose intended, mold designers should not agree to make a product drawing, and if they do, they must insist that it be signed by the customer as acceptable. This will eliminate any possible unpleasantness later on, if the product does not look or function as expected.

5.1.3   Are the Tolerances Reasonable?

Are the requested product tolerances feasible, in view of the size of the product and the plastic specified? This is sometimes overlooked when quoting. As we have seen in Section 4.10, while it is nearly always possible to make the mold parts accurately, to very close tolerances, this does not mean that the molded part will satisfy often unreasonable and unnecessary requests for close tolerances. If very close product tolerances are wanted, an experimental setup may be required to determine steel sizes, a process that can be very costly and time-consuming. This must be made clear before work is started. Note that in the case of very stringent tolerances, production (the actual molding) can become very expensive, requiring close inspection of the molded products and possibly causing many rejects.

5.1.4   What are the Cycle Times?

The designer should never guarantee cycle times and must make sure that the customer understands this. If the customer insists on any guarantee, it could require experimental work (test molds, remaking of mold parts, etc.), which could become very expensive. Any such anticipated costs should be brought to the attention of the customer, and added to the mold price. However, the designer should have some idea of the expected cycle, from past experience with similar products, or should try to get this information from someone with molding experience with such products.

5.1.5   What is the Expected Production?

The designer must be aware of the total production expected from the mold, and the expected life of it. There is a significant difference if the mold should be built for 1000, 100,000, 1,000,000, or 10,000,000 or more parts. This consideration will affect all aspects of a mold, from mold materials selection to many mold features selected by the designer.

It cannot be repeated often enough that the mold is the most important, but only one link in a chain of requirements to produce a molded product. The molder, or the final user, should not really be interested in the mold cost, but only in the cost of the molded product. It is the duty of the designer to advise the customer accordingly and build the most economical mold for the intended job.

The following is also a frequent scenario: A new widget is to be marketed. After a few hundred test samples, the customer estimates that during the next year he could sell 10,000 pieces. He does not yet know if the widget will be accepted at large. What size mold will be required? How will the mold cost, divided by this quantity, affect the cost of the widget? Obviously, because of the small quantity, the mold cost will be significant in this calculation. Also, because of the relatively small quantity, there may be only one cavity or at most 2 or 4 cavities required. This means low productivity, resulting in a higher molding cost. A simple cold runner system could be suitable and quite inexpensive. But what if the widget turns out to be a success and the required quantities increase to an estimated 1,000,000 over the next 3 years? The first mold probably will not be able to produce these quantities in time. This will then require a new, much different mold, with more cavities, a hot runner system, and so on—in short, a more complicated mold, which will cost much more but, despite the higher mold cost, will result in a much lower cost of the molded piece. Which is the better mold? They are both good, and each one is suitable for the specified requirement.

5.1.6 What are the Machine Specifications?

Before starting, the designer must know the machine or machines on which the mold is to operate.

5.1.6.1 Mechanical Features

(1) *Tie bar clearances and platen size*, front to back, top to bottom. Will the planned mold fit on the platens? In some cases it is all right to have the mold larger than these dimensions, it may even overhang the platens, as long as the cavities are located within the area between the tie bars. In some (today rare) cases, it may be necessary to pull one or both top tie bars to be able to install the mold. If this is required, the designer must find out if the planned machines have provisions for easy tie bar pulling.

(2) *Locating ring size, sprue bushing radius.* The locating ring centers the injection half of the mold on the stationary (or "hot") platen. The sprue bushing

radius must fit the injection nozzle radius. There are standards, but make sure you have the appropriate sizes. Some of the machines for which the mold is planned may have different sizes, so more than one locating ring (or an adaptor ring) and different sprue bushings may be required.

(3) *Mold mounting holes and slot pattern* (Euro, SPI, or other standard?). How will the mold be mounted on the platens? The best method is where the mold halves are directly screwed onto the platens, using standard mounting holes on the platens or clearance holes on the platens with threaded holes in the mold. With this method the full holding force of the screw is utilized. But this is often not possible, especially if the mold must fit several, different machines. In these cases, mold clamps are frequently used, with the clamp screws making use of standard mounting holes or slots in the platens. The disadvantage of this method is that only a portion of the holding force of the screw is utilized.

(4) *Quick mold change features*. There are a number of commercial and proprietary systems, and the designer must get the specifications to fit the system before starting to design the mold.

(5) *Machine ejector*. The ejector force is usually about 10% of the clamp force, which is sufficient for most molds, but there are cases where this is not enough. The mold may have to be equipped with additional ejection means, often built-in hydraulic or air actuators. The machine ejectors are always on the moving platen, but their size and pattern will vary according to the builder's standards (Euro, SPI, other standard?). If the mold will make use of the machine ejectors it is important to know their size and location when designing the ejection mechanism.

(6) *Shut height*. This is the total height of the mold, that is, the distance from the mounting face of the cavity half to the mounting face of the moving half. This distance must not be greater than the maximum distance of the platen surfaces of the machine when in fully closed position. The machine specifications indicate maximum and minimum shut height. If the laid-out shut height is too great, there are several ways to reduce it: (a) Investigate whether all the shown mold plates are really necessary. In some molds, for example, the mounting plate under an ejector box can be omitted, by fastening the mold to the machine using the mold parallels (see Fig. 7.3). (b) Reduce the thickness of one or more of the mold plates. (c) If neither is possible without compromising the quality (strength) of the mold, a different machine must be selected. This should be discussed with the molder before proceeding.

Conversely, if the shut height is too small, plate thicknesses can be increased, which is not always a good solution because it makes the mold unnecessarily heavy and adds cost to the mold. Some machines are equipped with Bolster plates, or bolster blocks, which are mounted on the moving platen in order to decrease the minimum shut height.

(7) *Clamp stroke*. In most machines, the mold clamp stroke is adjustable. For many molds, the suggested minimum stroke should be about 2.5 times the height of the product to ensure that the molded pieces have enough space to fall free between the mold halves during ejection; however, the stroke should not be less than about 150 mm (6 inches), so that the mold surfaces can be accessed for servicing while the mold is open. There are exceptions to these two suggested values, for special applications, particularly when using automatic (robotic) product removal methods, which are outside the scope of this book.

(8) *Ejector stroke*. This stroke is also adjustable, within the limits of the machine specifications. The designer must make sure that the available ejection stroke is large enough to push the products completely off the cores, in cases where little draft is specified, for example, when molding deep-draw containers. With good draft, it is usually not necessary to do more than push the products some short distance before they fall free, or before air-assist features will blow them away. There are again some exceptions, particularly with robotic product removal methods.

(9) *Clamping force*. The designer must make sure that the total projected areas of all cavities, plus the projected areas of any runner system in the same parting plane, multiplied by the estimated injection pressure, will not be greater than the available machine clamping force. As we have seen earlier, the estimated injection pressure depends on the ease of plastic flow (viscosity, temperature) and on the wall thickness of the product. In borderline cases, it is sometimes possible to change conditions, for example, in a very large product, by increasing the number of gates and placing them far apart; it may then be possible to use lower injection pressures, thereby requiring less clamp.

(10) *Auxiliary controls*. Some molds may require specially designed air circuits for air ejection or for air actuators. Is the machine equipped for such circuits, to be timed within the molding cycles? In some cases, hydraulically actuated side cores may be required. Has the machine a provision for timed core pulls?

5.1.6.2   Productivity Features

(1) *Shot size* (mass per shot). The total calculated or estimated shot size, that is, the total mass (weight) of the products coming from all cavities, plus the mass of the runner system (in the case of cold runners) should be within 30–90% of

the shot capacity of the machine. The shot capacity of a machine is given in g/shot of PS, with a specific gravity of about 1.05. The specific gravity of materials such as PE and PP is less (about 0.90 to 0.95); that is, the same mass will have a greater volume. Since shot size is rated in grams (or ounces) but is actually a volume (cross section of extruder barrel times the stroke of the extruder), the shot size of these materials will be less than for PS, by about 10%. These are only approximate figures; exact values should be checked with materials suppliers. What are the practical implications? If, for example, an 8-cavity mold is required to run in a specific machine, but its shot capacity is not large enough, it would not make sense to build it for this machine. This is especially important with cold runner molds, where the mass of the runner can add considerably to the mass of the sum of all molded parts, per shot. A machine could be well suited for a hot runner mold but be unsuited for a cold runner mold for the same number of cavities. (This is a major advantage of the hot runner system.)

(2) *Plasticizing capacity* (kilograms per hour). Plasticizing capacity is the amount (mass) of plastic a machine can plasticize per hour, that is, melt the cold plastic pellets into a melt of a specific temperature (and viscosity). Plasticizing capacity is usually given as mass for PS, in kilograms (pounds) per hour. Here, the same applies as with shot capacity. The actual mass of other materials, such as PE, PP, or any other, will be different, mostly smaller, sometimes greater. This should be carefully considered before starting. But, first, the designer must estimate the molding cycle, to find out how much plastic per hour will be required. Dividing 3600 (1 hour equals 3600 seconds) by the number of the seconds of the estimated cycle will give the number of shots per hour (N). Multiplying the total shot weight S (g/shot) calculated in (1) above, with the number of shots N per hour we find the total mass W_t in grams per hour required ($W_t = S \times N$). For best quality of the melt (and the molded piece), it is also suggested to use only between 30 and 90% of the rated plasticizing capacity. If W_t is more than the rated capacity, the machine can still be used but the cycle time will have to be lengthened; in other words, fewer shots per hour can be produced than the mold could yield with a suitable, larger size machine.

(3) *Injection speed* (grams injected into the mold per second). This is an important consideration when molding thin-walled products. Because of the narrow gap through which the plastic must flow within the cavity space, the injected plastic will cool rapidly when in contact with the cooled cavity and core walls. As the plastic cools, the gap narrows even more, making it more difficult to fill the mold. To overcome this condition, the melt and/or the mold temperatures could be increased so that the plastic will not freeze before filling the mold. However, this increase in temperature will also cause an increase in

the cooling cycle (and a lengthening of the molding cycle), resulting in a smaller output from the mold. This points to two areas for possible remedy: (1) The injection speed and (2) the injection pressure must be increased. But these two are interrelated. The higher the pressure, the faster the melt will be pushed through its paths, from the machine nozzle to the farthest corners of the cavity space. The problem is now that the injection speed depends on the speed with which the hydraulic injection cylinder is filled with pressure oil. Therefore, the speed of the injection cylinder depends on the hydraulic pump output—oil volume per second—entering the cylinder, but it also depends on the size of the associated hardware—hoses, valves, and so on—from the pump to the cylinder.

Most machines for conventional (not thin-wall) products are served sufficiently well by the output of the pump (and the motor driving it). However, the injection speeds required for thin-wall production require the cylinder to be filled more rapidly than what the pump alone can provide. To remedy this, the machine could be equipped with a much larger pump and motor, but in many cases this would be uneconomical or impractical. The preferred solution is to provide the machine injection system with an accumulator, which stores high-pressure oil during the time pressure oil is not used. Additional valving and other hardware is required, which is often sold as an "option" with the machine, called an accumulator package. The accumulator releases the stored high-pressure oil together with the pump output into the cylinder when required for injection. The designer will need to recognize when an accumulator package is necessary for the product for which the mold is to be designed, and must discuss this with the molder to make sure the right machine is available to run the mold.

5.1.6.3   Additional Requirements for Some Molds

(1) *Pressure air.* Some molds require air pressure for their operation. In general, the designer should be aware that compressed air, especially in large volumes, can be very expensive, especially if it is left to blow for any length of time.

- *Blow downs* (air jets or air curtains) are often used to assist the products to rapidly clear the molding area. There are several commercial air jets on the market with low consumption of pressure air. Their initial cost is paid back rapidly by savings from wasted air volume.
- *Air-operated actuators.* The air volume used is usually small, compared with a blow down. There could be problems with controlling the speed

and uniform motion of air actuators, but they are simple and inexpensive.

- *Air required for air ejection*, which is usually activated on demand, for a very short time. Most of the time, the actuation time is controlled from the machine control panel. The designer must make sure that the intended machine is equipped with sufficient controls and hardware (timers, valves, and large enough supply lines). It may be even necessary to add pump capacity, for the added volume of air that will be required for the planned mold. If much air is needed for short blasts, one or several suitable accumulators could be installed near or even on the mold. This is similar to the hydraulic accumulators cited in Section 5.1.6.2 (3).

Where pressure air comes into contact with the molded products, for example, in blow downs or in air ejection, the air must be filtered from any oil residues, water (always present in air lines), and so on, before reaching the outlets in or at the mold, to prevent contamination of the products if they are used for food or pharmaceutical purposes. (Unfortunately, most air actuators require lubricated air, unless their seals are selected for dry air.) A low-pressure, high-volume blower with its air intake from the shop environment, or better yet, from within an enclosure built around the molding machine when special "clean room" requirements are specified, is a preferred solution to ensure that there is no oil or water contamination in the air as it comes into contact with the plastic products. In many cases, such blower can be directly mounted on the top of the mold. Another advantage is that the power consumption of this type blower is low, on the order of 0.2 kW (1/4 hp) or less, and does not require timing or valving.

(2) *Auxiliary hydraulic supply.* For some operations, compressed air may be not suitable. (a) Air cylinders are often jerky in their operation, especially with long strokes. (b) In cases where several air cylinders actuate one large mold member, the forces can be uneven and the member can jam. (c) In most molding shops the compressed air pressure is fairly low, usually about 600 kPa (80 psi), and rarely 900 kPa (120 psi), so large air actuators are needed to produce large forces. It could be difficult to accommodate sufficiently large cylinders within the available mold space, or even outside the mold. In all these cases, the much more powerful hydraulic cylinders would be an alternative. The hydraulic pressure could be taken from the machine system with a pressure reducing valve, and by providing the necessary safety measures to protect against the very high pressures in that system. A preferred method, however, is to use an auxiliary power supply, usually at a system pressure of about 3,500 kPa

(500 psi). This is much safer and requires much less expensive hardware (valves, hoses, etc.) than that for higher pressure. The motion of hydraulic operators is smooth and the speed can be well controlled.

Two points of caution, though. Hydraulic oil (with some special, expensive, exceptions) is highly flammable and there is always the danger of leaks, especially if the leaks were to occur near heated areas of the mold, as, for example, near a hot runner system. Also, products used in the food or pharmaceutical industry could be contaminated by the oil; this is usually specified as not allowed.

(3) *Cooling water supply.* This is a very important area of concern. There is not much sense in designing the mold with very sophisticated cooling circuitry if the cooling water supply is insufficient in temperature, volume, and pressure. An individual chiller unit may be the answer if the plant supply is too small or has not enough pressure. It is also important that the coolant is clean, that is, with a minimum of minerals or dirt, and is not corrosive. Dirty coolant could gradually plug the water circuits or coat the channel walls with a poor heat conducting layer of dirt and lime, thus reducing the cooling efficiency, and could require frequent cleaning of the coolant channels if the mold is expected to maintain high productivity. Corrosive action of the coolant could attack and eat away the mold steels; rust creates insulating layers similar to lime and dirt deposits. It is always good policy for the designer to check with the molder to ensure that there are no such problems with the water supply, and to specify that only clean, noncorrosive coolant is used with the mold. See *ME*, Chapter 13.

(4) *Electric power and controls.* The electric power supply in North America and in most developed countries is usually sufficiently stable and uninterrupted, except during natural catastrophes, and of not much concern to the designer. This is not the case in developing countries, where power interruptions occur frequently; the effects of such interruptions on the operation of a mold may cause concern. Typically, in the case of a power failure, a machine using a cold runner mold will just stop, but can resume work as soon as the plastic is again up to molding temperature. However, in a hot runner mold the melt will freeze in the manifold and nozzles and it may take much more time to restart (in small molds between 15 and 30 minutes). The expected savings through using a hot-runner mold may become an illusion. The controls (breakers, heat controllers) available to operate a mold on a specific machine must be discussed with the molder when designing a mold that will require additional heat controls; typically, such controls are required for hot runner molds. For safety reasons, heaters in molds are rated at 230 VAC or less, and the power consumption may be from as low as 40 W per heater, such as in some nozzle heaters, and up to several thousand watts in hot runner manifold heaters.

Since heaters are often bundled in parallel and operated by designated controls, it is important to ensure that adequately sized circuit breakers and so on are available; some can be controlled with time-percentage controllers or variable (voltage) transformers, whereas some will need thermocouples and heat controllers.

5.2   Start of Mold Design

Now that all our preliminaries are clear, the designer must decide what kind of mold should be designed. With the expected production in mind, the most suitable, that is, the most economical mold for the job must be selected. As was already stated earlier, a very expensive mold intended for high productivity will not necessarily be the best choice. The designer must always find the most cost-effective mold, that is, the mold that will result in the lowest cost of the product.

5.2.1   Mold Shoes

A mold shoe (sometimes also called "chase") is the total of all mold plates making up the mold, including screws and alignment features, but not including the stack, which is the arrangement of all mold parts that touch the injected plastic, typically, the cavities, cores, any inserts in either of them, ejectors, strippers, side cores, and so on. In simple molds (not necessarily low-production molds) the cavities and cores can be machined right into the mold plates. A decision on which way to proceed with the mold shoe should be made only after the product drawing is carefully studied, and never losing sight of the expected productivity of the mold. There are several choices for the designer.

5.2.1.1   No Mold Shoe Used

The mold may consist of only one plate for the cavity and another plate for the core, with both cavity and core machined right into these plates. Ejection is facilitated by air valves built directly into the core plate. The entire mold, then, consists essentially of only two parts, plus alignment features and air valves. Cooling channels are built right into the plates.

5.2.1.2   Standard Mold Shoes

Mold shoes can be bought from mold maker supply houses (DME, Hasko, etc.) from a large selection of standard sizes, with or without leader pin alignment, and with or without ejector plates. All plates are machined and ground square, and are ready for adding the required mold features. Many mold shops prefer to buy these ready-made mold parts, and rather specialize in the making of the stacks and doing the final mold assembly. These plates are usually available in several qualities of steel:

(1) The first type is an inexpensive, "mild steel," which is soft, with low strength, and little wear resistance. It is suitable only where the expected forces and wear in the mold are small enough that the steel will not be damaged, for example, by the clamping pressure on a too small P/L, or by the hobbing effect, which is the pushing of a supported small insert into a mild steel backing. Also, since mild steels have a low tensile (and compressive) strength, they may permanently deform if loaded beyond their yield point.

(2) Another common steel supplied is a type of "machinery steel," typically a steel called P20 or P20PQ (plastic mold quality). It is treated to a Rockwell hardness of approximately Rc 30–35; P20PQ is produced especially clean, that is without dirt enclosures, which could be detrimental if they appear on a molding surface. These plates are more expensive but cost much less than so-called mold steels; they are very suitable for cutting the cavity or core right into the plates. This is of special advantage for large products where the cost of mold steel would be very high. Mold steel is always supplied very soft (about the same as mild steel), for easy machining; it must be hardened and ground after machining, which represents an additional, considerable expense. In high-quality molds, the stack parts are usually made from steels such as P20PQ for large products, and from mold steels for smaller products.

(3) In high-quality molds, also, both the mold shoe and the stack parts are made from stainless steels (SS). The larger mold parts are then machined from prehardened steel, and smaller parts from SS mold steels. This is helpful in humid climates to prevent rusting of the mold shoe, or where the plastic is corrosive and could attack the stack parts. The higher material cost can often be justified with savings in mold maintenance. Note that for corrosive plastics (e.g., PVC) the stack parts made from regular mold steels must be chrome plated, which is expensive and requires additional maintenance. Mild steel plates and P20 plates can be protected against rust by a relatively low-cost electroless nickel coating, or by just oiling well after use, before storing the mold. (More about mold steels in Chapter 9.)

5.2.1.3 Home-Made Mold Shoes

The mold shoes can be made in-house from raw steel plates, which the mold maker can buy from the steel mills or dealers. The mold maker may keep certain plate sizes and thicknesses in stock, and cut and machine them to size as needed. This requires much plant space, heavy lifting and storage equipment, and accurate milling and grinding machines. It is an economic decision that may be different from shop to shop—whether to make the plates or buy them as standard plates or as complete mold shoes. In any case, the same choices of steel apply. Note that often, such in-house made mold shoes or plates are built to the dimensions listed as standard parts by the hardware suppliers.

5.2.1.4 Special Mold Shoes

This applies mostly to special molds for which no suitable standard sizes are commercially available, and to very high production molds, where the mold shoe is built around the stacks, and optimum layouts are used for all mold features, rather than the stacks being fit into the space available in standard molds. Some mold makers specializing in certain areas (preform molds, unscrewing molds, etc.) create their own mold shoe standards. In high-production molds, the mold shoe too is usually made from prehardened steel. Also, often prehardened stainless steel is used for such molds.

5.2.1.5 Universal Mold Shoes

For low production and relatively small products, a universal mold shoe offers another solution for making a relatively low-cost mold. Universal mold shoes are essentially standard size chases that are constructed so that different stacks can be easily mounted into them. The mold maker concentrates on making the stacks, in sizes and to rules specified by the maker of these universal mold shoes. Mold features such as runners, cooling, and ejection are usually not as efficient as in a mold specifically designed for a product, and the mold will not cycle as fast; but, for the small quantities required this is no problem, and the mold is much less expensive than a complete mold. This makes a lot of sense, especially if a large number of such "inserts" are used or foreseen.

5.2.1.6   Mold Hardware

Hardware items include leader pins, bushings, ejector pins and sleeves, screws, and many other mold parts that are required for the mold. They are all listed in catalogues issued by the various mold maker supply houses; they are mass produced, using high-quality materials, and machined to very close tolerances. It is always more economical to buy these parts rather than to attempt to make them in-house. Also, a good mold designer will never modify these products, with only one exception: the cutting to length of the ejector pins and sleeves. A diameter should never be modified; a way can always be found to make the design use a standard size diameter. Also, screws used in molds must never be modified, not even their length; there is always a way to make the design use a standard size, often by just changing the depth of a counter bore for the screw head. Any modification of a screw will reduce its strength; a modified screw is also difficult to replace in the field. Because screws should be tightened to about 60–70% of their yield strength, in good maintenance procedures all screws should be replaced every time the mold undergoes a major overhaul.

5.2.2   Mold Drawings

5.2.2.1   Assembly and Detail Drawings

The purpose of the assembly drawing (including the Bill of Materials discussed later) is to convey the intentions of the designer to the people involved in purchasing hardware and materials, assembling the mold, and, finally, operating the mold. The assembly drawing of the mold must contain all pertinent information, given in plan and section views and in notes, which are used to explain where the drawings alone could be ambivalent or misinterpreted. Once the assembly drawing is finished, there must be no doubt left about how the mold is to be built and operated. Today, most mold makers depend on machinists specialized in their trade, such as lathe, milling machine, EDM, or other machine tool operators. These machinists need detail drawings, complete with tolerances and, if deemed necessary, other instructions such as hardness, plating, and finishing. These detail drawings are prepared from the assembly drawings.

5.2.2.2   How Many Drawings and Views?

This question is frequently asked. The answer is simple: enough to make sure that there is no possible misreading of a drawing. Too few views (or sections) means that in the best case, the machinist will interrupt his work to come and ask for explanations, which costs in lost time. In the worst case, the machinist will not ask, but will proceed in the wrong direction. This could become very expensive if an incorrectly made piece is not discovered until it reaches assembly, and then has to be remade, or it could cause major interruptions until a solution is found to use and repair the wrong part. Sometimes other mold parts have to be altered to make it possible to use an incorrectly made but expensive part. On the other hand, too many, often unnecessary views make more work for the detailer and can be confusing for the user of the drawings.

5.2.2.3   Arrangement of Views

Most molds are laid out by starting from a (significant) cross section and then drawing to the right of it (as the mold would be when mounted in the molding machine) a view into the cavity half of the mold, that is, into the injection side (see Fig. 5.1). The assembly drawing should show above this view words such as "Plan view into cavities."

On the left side of the section view, the core half is shown, as if looking into the direction of the core and the moving platen. The assembly drawing should

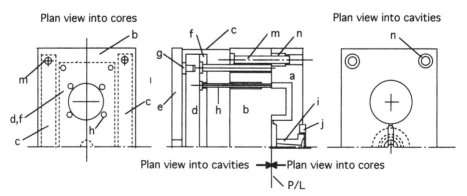

Figure 5.1   Arrangement of mold drawing layout: (a) cavity (plate), (b) core (plate), (c) parallels, (d) ejector plate, (e) mounting plate, (f) ejector retainer plate, (g) stop button, (h) ejector pin, (i) sprue bushing, (j) locating ring, (m) leader pin, (n) leader pin bushing.

show above this view words such as "Plan view into cores." The plan view drawings are made so that we see the parting line (plane) as visible, and all plates and mold features behind it as invisible lines. Additional full or partial cross sections and/or plan views should be added (usually on separate sheets) only when they can add information to the views already shown. Remember, the designer is in the business of designing molds, not making pretty pictures. However, the drawing must always be drawn to scale, so that the various parts can be seen in proper proportion. "To scale" in this context means to draw to a selected, set ratio, preferably "to size" ($1 : 1$), or if this is not practical, in case of large products, smaller, often $1 : 2$, or even $1 : 5$, or larger, often $2 : 1$, $5 : 1$ or $10 : 1$.

5.2.2.4   Notes on Drawings

Whenever it is impossible or cumbersome to specify some important information by using standard drawing techniques, a note should be added to express in concise but clear words what is intended. The note should be short, but not so short that it could be open to misinterpretation. Also, the drawing should not be cluttered with too many notes, and must show clearly what the notes apply to.

5.2.2.5   Additional Information on the Drawings

For more complicated molds, it is good practice to show also, on another sheet if necessary, separate schematic views of (1) all coolant circuits, (2) any air and hydraulic circuits, (3) any special electric circuits, and (4) a sequence of operation of the various mold functions, for example, at what point in the cycle ejection starts, and when air should be activated. This can have legal implications: complete and correct information will protect the designer from any possible future litigation, in case of an accident caused by the mold not being installed and operated as recommended by the designer. Note that the drawings are part of the job and must be shipped to the customer together with the mold.

5.2.3   The Stack Layout

By now, the designer will probably have decided what type of mold should be built, guided by the possibilities discussed in Section 5.2.1. This does not necessarily mean that the designer is bound by this early decision. It may become necessary to reconsider as the design progresses. The designer must always keep an open mind and be ready to scrap an earlier idea for a better one. The more time spent on thinking and rethinking the problems at this time, the more successful will be the final result; this will save time and money in the long run.

5.2.3.1   Significant Cross Section

Which type of mold shoe will be finally selected for the job is, at this point, of secondary importance. The designer must now start with showing, to scale, a significant cross section of the product. This means the section that shows all the areas that must be considered when designing the stack. (If more than one significant feature cannot be shown in the main cross section, additional—or partial—sections may have to be shown.)

The cross sections will now be examined and a number of questions will have to be asked, step by step.

5.2.3.2   Will the Product Slide (Pull) out of the Cavity?

This point should be investigated first, because it will determine the complexity of the cavity.

(1) In the case of a simple, cup-shaped product, there is usually no problem.

(2) Holes (cutouts) or projections in the side wall of the product may require special attention: will it be necessary to provide side cores? If side cores, should there be one for each hole or one for a group of holes or projections? Should the cavity have a complete side wall moving? In the case of beverage crates, all four walls may have to move, which will create four vertical split (parting) lines. All this will considerably

increase the complexity of the mold and increase the space required for each cavity and for the stack in general.

(3)   Are there other projections in the side wall of the product? If they are deep, they will probably be considered like holes. If they are shallow (for example, engraved printing or ornamentation), it will depend on the draft angle of the side wall and the plastic injected. In some cases, typically with draft angles over 5°, shallow engraving could pull out of the cavity, provided there are features (such as undercuts) on the core to ensure that there is enough force on the molded piece to pull it out of the cavity. Cases like this should be discussed with the designer of the product for which the mold will be built. There may be a good chance that the product design could be slightly changed such that side cores are not necessary at all, thereby saving considerable expense.

(4)   Other possibilities can be considered, especially for large openings in the side walls, as, for example, the large cutouts in the sides of a typical, large laundry basket, where the cavity and the core can meet at an angle and produce additional, small parting lines, but not require side cores or split cavities.

(5)   The most common case is where the cavities split into two halves, creating two vertical split lines.

(6)   There may not always be enough space for the long side motions required for two splits, and the cavity will split into four sections; this is common with pail molds where four moving side cores are wedged within the cavity block walls to contain the outward forces of the side cores. Note that in all cases where side cores are used, they must be preloaded and backed up against the forces generated by the injection pressure.

5.2.3.3   Will the Product Eject Easily from the Core?

Are there any raised portions inside the product that would be molded in severe undercuts in the core and prevent the product from being ejected easily?

(1) *Snap* (Fig. 5.2). A frequently used undercut is a snap feature, which is a (usually circular) rim inside the product, shaped to snap over a similar extension in a matching product, for example, the lid over a can. Provided the shape of the snap rim (its cross section—tapered and/or rounded sufficiently—and the total circumferential length) is suitably designed for ejection, there is no problem, and a stripper ring will easily eject the product by forcing the rim to expand while

Figure 5.2   (*Left*) Section through a cap with snap; (*Right*) example of 4-section snap.

ejecting ("stripping"). Stripping with ejector pins, located at some strategic points, may also be used for stripping, in nonround products. To make the snap ring easier to stretch and to come off the core without breaking, it can sometimes be broken down into several sections, so that there will be, for example, four sections, each covering about 60–70° of the circumference, instead of covering the whole 360°. Of course, customer's approval must be secured before making such a change. Note that this is more difficult to machine.

(2) *Internal threads*. If the threads are designed suitable for stripping, they can be stripped from the core like the snap rim described above. It is better if there is not more than one complete thread (360°). Multiple threads may cause damage to the molded thread projections as they are dragged over the depressions for the successive threads in the core during ejection. (In many cases, one thread may be strong enough for its intended purpose.) It must also be understood that there is a relationship between the amount the plastic that is stretched radially and circumferentially. For a certain cross section of the snap rim or thread, and if the product is small, there may not be enough length in the circumference to stretch, and the product will tear.

(3) *Unscrewing*. In some cases, the product must be unscrewed from the core, which means a much more complicated (and expensive) mold. There are several moldmakers specializing in unscrewing molds, using standard design mold shoes and stacks, thereby reducing the cost of such molds.

(4) *Undercuts*. Undercuts are used to hold the product on the core, to ensure proper ejection, especially in cases where the product is designed so that it could stay in the cavity while the mold opens, often held by the vacuum between product and cavity wall. With many products, there are enough "vertical" surfaces in the core, such as slots for ribs, or specially shaped slots and holes as often required in technical products, to hold the product on the core side; the same is true for cup-shaped products with little side draft, where the product will tend to shrink tightly onto the core. If this is not enough to hold the product on the core, judiciously designed and placed undercuts should be specified at the time of designing; do not leave it to the molder to add undercuts after the mold is in operation and causes ejection problems. The proper location for these

undercuts (which are usually not specified by the product designer) is (a) near ejectors, or (b) preferably, especially with hot runner molds, near the tip of the core where the undercuts are more effective because the bottom of the product is stiffer.

(5) *Two-stage ejection.* Two-stage ejection (Fig. 5.3) may be a solution for some, somewhat larger undercuts inside the core. (See *ME*, Chapter 12) This a more expensive design of the core and the ejection mechanism, but it is frequently used in products that require a snap design inside the product. It is often used for overcaps for spray bottles that are produced in really large quantities. The cooling is less efficient, but it is a well-accepted and reliable design.

(6) *Deep projection inside the product.* This feature often requires very complicated core design, possibly with moving, retractable core sections, or "collapsible cores." Both systems are expensive, difficult to build, and hard to maintain in operation; they are also usually difficult to cool adequately, and thus run much slower than a comparable mold without these features. (See *ME*, Chapter 12)

5.2.3.4   Establishing the Parting Line

(1) *Primary parting line.* Before proceeding, the location of the dividing plane (parting line, P/L) between cavity and core must be decided. In cup-shaped products, this is usually simple: it is at the widest portion of the product. As stated earlier, a straight P/L is easiest to produce, preferably, but not necessarily, at right angles to the direction of the mold opening, that is, the axis

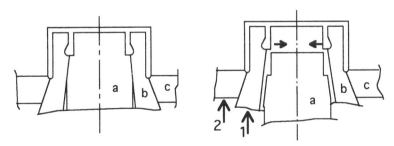

Figure 5.3   Schematic of 2-stage ejection: (a) core, (b) sleeve, (c) stripper ring. During ejection, first (1) and (2) move together so that the core can slide out, then the stripper moves up to push the product off the sleeve while the projection on the inner sleeve moves inwards, as shown by the small arrows.

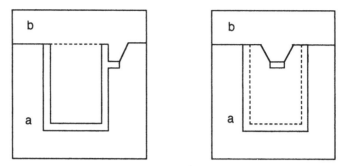

Figure 5.4   Schematic of mug with handle, showing offset parting line: (a) cavity, (b) core.

of the mold. An offset (or stepped) P/L is sometimes required, due to the shape of the rim. It is also, occasionally, used for molding a large projection, for example, a simple handle of a mug, on the outside of the product; such an offset P/L is preferable to a side core, which would be much more expensive to build (see Fig. 5.4).

(2) *Split cavities or side cores.* At this time the designer must also determine if the cavity needs to be split and where the split lines will be located, or if side cores will be required. Usually, but not always, the split lines are parallel to the axis of the mold, and side cores at right angles to it. Both split cavities and side cores need backing up and preload against the forces created by the injection pressure, and some method of operating mechanism, which will also require space in the mold. Operating mechanisms can be angle pins (horn pins), or rollers in tracks, both of which translate the opening motion of the mold into sideways motion; they could also be timed, hydraulic actuators independent of

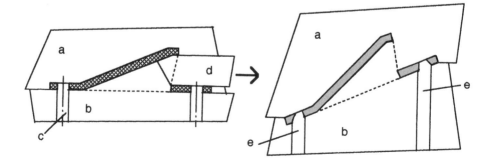

Figure 5.5   Example of a louver mold: (a) cavity, (b) core, (c) round core pin, (d) side core, (e) core pin with shaped tip.

the clamp motion. The designer should also consider if a better mold layout could be achieved by turning the product slightly, to achieve with a straight (up-and-down) mold what would otherwise require side cores (see Fig 5.5).

In some cases, rotating the product 90° could also result in a better mold, as shown schematically in Fig. 5.6. The right schematic shows the normally expected mold layout, with the center line of the product parallel to the mold axis. Because of the outside shape (e.g., deep engravings or projections) the cavity will have to be split; the projected area (at right angles to the axis) of the product is very small compared to the projected area of the sides of the product. Therefore, the side cores will see considerably larger forces *Fs* at right angles to the mold axis. These forces must be adequately backed up and preload provided to prevent the splits from cracking open during injection. (See also Section 5.3.)

These backups, especially for large splits, can result in a very bulky mold. But by turning the product by 90° (left schematic, in Fig. 5.6) the primary P/L replaces the split line, and the cavity and core halves are clamped by the machine clamping force *Fc*. The core must now be withdrawn sideways; it will have a much smaller area exposed to the injection pressure and will need much less backing-up force, but the stroke of such side cores will probably be much greater than the stroke of the split cavities. This could be an undesirable feature, but is often preferred to the alternative of split cavities. Only by laying out to scale these alternatives, at this time of the design process, will the designer be able to determine which is better for the contemplated mold and how to proceed. Note that in the position shown in Fig. 5.6, the open end of the product is on top and the product is ejected downward from the (side) core and can fall unhindered. If the core has sufficient draft, air pressure alone could be sufficient to eject the product from the core, which would make for a much simpler mold.

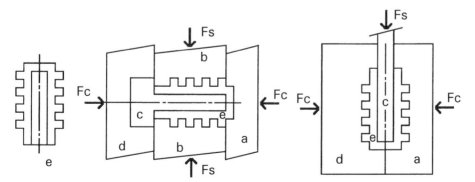

Figure 5.6 Illustration of a product and two possibilities of mold layout: (a) cavity, (b) splits, (c) core, (d) core backing plate, (e) product, (*Fc*) clamp force, (*Fs*) force on splits.

5.2.3.5   Is the Cavity Balanced?

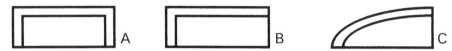

Figure 5.7   Schematic of cross section of (A) cup-shaped product; (B), (C) open-sided product.

Figure 5.7 shows the elements present in every cavity shape. Within (A), the cavity pressures are balanced; in (B) and (C), the cavity is imbalanced. It does not matter if the side walls are at right angles to the internal pressure. There is always a component of the pressure that will press in the direction at right angles to the mold axis. As can be seen in Fig. 5.8, on the left, the pressures inside the cavity push to the left and the right by the same amount, and there will be no force to move the core relative to the cavity. The cavity is therefore balanced.

In the drawing on the right, the pressure p within the cavity tries to separate the cavity and the core by pushing the cavity to the left and the core to the right, as indicated with heavy arrows. This must be taken into account when designing a mold with imbalance in the cavities. The force trying to separate cavity and core can be balanced by placing a second, similar stack near the first one so that the forces are pushing in opposite directions. Failing this, there could be one pair of wedges (similar to a wedge lock) located so that the imbalance is taken up there. If this is not done, the forces of the imbalance will have to be taken up by the leader pins and bushings, which may not be strong enough in some cases, and wear rapidly.

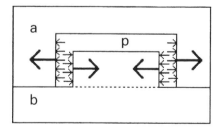 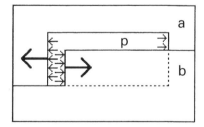

Figure 5.8   Schematic illustration of (*left*) a balanced and (*right*) an imbalanced mold: (a) cavity, (b) core, (p) internal injection pressure.

5.2.3.6   Determining the Method of Cavity Construction

(1) *Cavity and/or core are cut right into the mold plates.* This would make the simplest mold. Some molds have only one or a few cavities cut into the mold

plate, but the cores are usually separate from and mounted in or on the core plate. For practical reasons, one-piece cavity and core plates are often selected for single-cavity molds, mainly for very large products, but there have been molds like this built for smaller products, with more cavities. The problem is to provide the necessary accuracy of machining, especially in the absence of sufficiently large, accurate machine tools. The mold could consist of fewer parts, and if there are no foreseeable problems with ejection, cooling, and mold life, it is a good method, especially for very large molds. The mold steel selected should be of "mold quality," prehardened; a typical mold steel is P20PQ, or stainless steel, prehardened. Such cavities and cores may still require inserts (usually pins) whenever small holes and so on in the product would require delicate projections in the molding surface. To machine such projections from the solid steel, while possible, would be very costly to repair if they should be damaged.

(2) *Composite cavities and cores.* Individual, solid cavities are cut from mold steel, with inserts as needed; this is the most common design. These cavities are then either mounted on top of the cavity plate or inserted into it. Cavities can also consist of an assembly of separate pieces, arranged to form the cavity assembly. If the outside of such an assembly is a (not necessarily round, but sufficiently strong) ring (or "chase") into which the inserts are placed, this assembled cavity can be treated as a solid cavity and mounted on top of the cavity plate or be inserted into it. In some cases, the inserts are directly placed inside the cavity plate, without the need for a surrounding ring. Note that the forces from the injection pressure on the sides of the cavities are considerable, especially if the projected area at right angles to the mold axis is large, that is, where the product is deep. These forces will tend to loosen the inserts and can create gaps between them or between inserts and cavity plate, where plastic can flash into. Properly designed, such inserts must be pressed into their chase (or into the mold plate) to create a preload larger than the expected side forces.

5.2.3.7   Determining the Total Area of the Stack

The total area of the stack is the total of the space of the cavity (including the ring discussed above, which may also include cooling channels) plus the area (space) of any added features that may be required, such as side core components outside the cavity or core, plus space for their motion, actuation, and the backup. It can be seen that this total space can be much larger than the cavity by itself, and will determine the size of the mold and affect the cavity layout. It is easily understood that a mold without side cores requires much less space, and a much smaller layout, for the same number of cavities.

5.2.3.8   Determining the Core Construction

Cores may require quite a number of inserts and even moving parts; the injection pressure is usually of little concern (except in some special cases) because this pressure tends to compress the core from all directions rather than expand it as it does the cavity, and is resisted by the compressive strength of the core material. There is one serious problem, though—the "core shift," especially with long slender cores, when the flow and the pressure of the inrushing plastic can deflect a core, resulting in uneven wall thicknesses around the core. This is mostly of concern with thin-walled products, which require higher injection pressures, and where uneven wall thicknesses can create differential pressures on opposing sides of a core, thereby creating forces that deflect (bend) the core during injection. Such deflected cores return to their original shape as soon as the product is ejected, but by that time it already has uneven walls. Problems like this can sometimes be solved by supporting the tip of the core in a matching hole in the cavity when the mold is closed or by some other, often patented methods. Core shift can also be affected by the location of the gate; multiple gates are sometimes a solution. (See *ME*, Chapter 10.)

Cores are usually mounted on top of the core plate, either solidly (the most common method) or floating, which is better, but more expensive; they are rarely inserted into the core plate.

5.2.4   Selection of a Suitable Runner System

We must now consider how the plastic will be channeled from the machine nozzle to the cavity space. This could have been specified with the job order, but, nevertheless, we should understand the various systems and where they are most appropriate.

5.2.4.1   Cold Runner, Single-Cavity Molds

The arrangement, as shown in Fig. 5.9, is simple, effective, and good for very large products, but is also often used for smaller ones. The disadvantage is that the gate is large and must be cut or even machined if appearance is important.

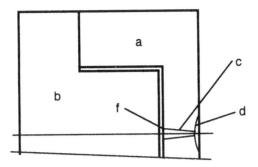

Figure 5.9   Schematic of (large) single-cavity mold: (a) cavity, (b) core, (c) sprue, (d) nozzle seat, (f) gate.

5.2.4.2   Cold Runner, 2-Plate Molds

The mold on the left in Fig. 5.10 has only one P/L. An edge-gated arrangement is shown. The products and the runners stay together when ejected and must be separated after molding. There are other methods of gating, some of which are self-degating as the mold opens, but products and runners are still mixed together and require separation. More about gates in *ME*, Chapter 10.

The *advantages* of this system are (1) simplicity and (2) low cost. Also, (3) color changes are easy, and (4) the system is not sensitive to dirt in the plastic. If a gate is blocked, it is clean again after the runner is ejected.

The *disadvantages* are (1) these molds usually have longer molding cycles because of the longer time required to cool the often large runners. (2) The mass

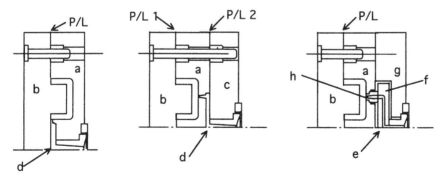

Figure 5.10   Schematic illustrations of (*left*) a 2-plate mold, (*center*) a 3-plate mold, and (*right*) a hot-runner mold: (a) cavity plate, (b) core plate, (c) third plate, (d) cold runner, (e) hot runner, (f) hot runner manifold, (g) hot runner backing plate, (h) nozzle. P/L, parting line.

of the runners must be added to the mass to be plasticized for the products, therefore, energy is wasted, first in plasticizing, then in cooling. In some cases, the mass of the runners is as great as the mass of the products, or even greater. (3) Although in many cases the runners can be reused, this requires more handling (costs), energy is needed for regrinding, and there is always a danger of contamination of the plastic. Also, losses of plastic (maybe 10% of the scrap) in the course of this process are unavoidable. Even so, 2-plate molds are used in the vast majority of multicavity molds.

5.2.4.3   Cold Runner, 3-Plate Molds

Three-plate molds are also cold runner molds, but the system is inherently self-degating. The mold in the center of Fig. 5.10 has *two* P/Ls. As the clamp opens, first, the cavity plate travels with the moving mold half; as soon as the cavity plate has reached a limited distance the moving mold half (the cores, with the products still on them) continues to move away from the cavity plate and the products can be ejected, after P/L 1 is opened. At the moment when the product, still on the cores, start pulling out of the cavity, the plastic in the gates is severed. Then, by some more or less complicated mechanism, P/L 2 separates and permits the ejection of the runner system in a separate plane.

Advantages: (1) The products can be center gated, or gated anywhere on the top surface. (2) Due to the absence of runners in the P/L, the cavities can be closer together and more cavities can be placed in a mold of comparable size; see the difference between left and center (or right) illustrations in Fig. 5.10. (3) The gate vestige is usually very small, with excellent appearance. (4) Color changes are easy. (5) The system is not sensitive to dirt in the plastic. If a gate is blocked, it is clean again after the runner is ejected.

Disadvantages: (1) Three-plate molds are much more complicated and expensive. (2) It is very difficult to guarantee 100% automatic ejection of the runner system. There are numerous systems, with links, chains, air actuators, and so on to provide the necessary motions. (3) With 3-plate molds, too, the runner mass can be greater than the total mass of the products; the same comments regarding productivity apply as for 2-plate molds.

This system is often used for very small products, such as screw caps, overcaps, and so on, which should be center gated for best molding flow, such as in containers, and where the appearance of the top surface of the molded piece is important.

5.2.4.4   Hot Runner (HR) Molds

In the system on the right in Fig. 5.10, the plastic (melt) is kept hot, on its way from the machine nozzle to the gate. Heaters in the sprue bushing, the HR manifold, and (usually) the HR nozzles (which terminate at the gate) ensure that the plastic stays at the required temperature. (Note that it is not the purpose of the hot runner system to add to or regulate the melt temperature, but just to keep it as it comes out of the machine nozzle.) When the mold stops operating, that is, when the power is off, the plastic in the hot runner will freeze. To make the plastic hot again to restart the mold in a reasonable time (15–30 minutes), the heaters must be strong. But during operation of the mold, the heat requirements are small, especially with a well-designed system with a minimum of heat losses to the surrounding, cooled plates. Some molds, once "on cycle," require no heat at all or as little a 5% of the rated heater capacity of the hot runner system. The main design problems in hot runner molds are the gate shape, the temperature profile around the gate, and the materials selection.

(1) *Open gates* depend on their size and shape and on the operating pressure and temperature of the plastic. At the end of the injection stroke, the gate must freeze sufficiently to stop the plastic from drooling into the cavity while the mold is open for ejection of the products. When the mold recloses, the injection pressure must push the frozen "plug" of plastic out of the gate into the cavity space and thereby permit the plastic to flow again.

(2) *Valved gates* are closed and opened by mechanical (or electrical) means, as timed. This requires more mechanisms and controls, thus adding to the mold cost. The size of the gates can be much larger than with open gates, which, in some cases, can be very important for the filling of the cavity spaces; it also reduces the sensitivity to dirt, because dirt can more easily pass through a large opening. Larger gates are also of advantage for materials that are sensitive to high stresses.

Advantages of hot runner molds: (1) The cavity spacing can be similar to a 3-plate mold, that is, closer, making good use of the available space. (2) The mold output can be greater since all material that is plasticized is used to produce products. (3) There is no need for regrinding, except for scrap during start-up.

Disadvantages: (1) Higher mold cost (but not much different from a 3-plate mold. (2) Difficult color changes. The plastic within the hot runner system must be completely clean before a new color can be used. A measure of a good hot runner system is the number of shots required to change from a darker color to a lighter one. A good HR will do this in about 15 shots, after clean, new color is coming from the injection unit. (3) Very sensitive to dirt in the plastic. If there is a gate blocked by dirt, the nozzles must be accessed for cleaning, which may take anywhere from half an hour to a day; it may even be necessary to remove the mold from the machine. A good mold design makes sure that this cleaning can be easily performed while the mold is in the machine. (4) Cost of plastic. The sensitivity to dirt also suggests that the molder should use virgin plastics rather than regrinds, which are more likely to be contaminated. This will affect the cost of the product. (5) With today's technology, there are still problems to mold very small products, because of the long residence time of the plastic in the runner system, which, if too long, causes the plastic to degrade.

5.2.4.5   Cold and Hot Runner Molds, in Combination

Combinations of hot and cold runner molds are usually selected for cases where cold runner molds (edge or center gated) would have advantages over hot runner molds. This is done sometimes for very small products, to avoid excessive residence time in the HR system, or for very large products, to reduce the pressure drop from machine nozzle to the gates, especially if the distances from nozzle to gates are very large.

Two typical examples are shown in Fig. 5.11. Many such combinations are possible.

(1) For multicavity molds (Fig. 5.11, *left*), the runner system could become quite large. To prevent large pressure drops while avoiding unnecessary large masses of plastic, the runner channels taper down from a heavy cross section where the plastic enters the mold at the sprue, becoming gradually smaller every

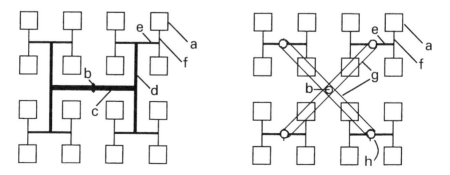

Figure 5.11   Schematic of a 16-cavity mold: (*Left*) common cold runner mold; (*Right*) combination hot and cold runner mold. (a) Cavity, (b) sprue, (c–f) cold runners, (g) hot runner.

time the runner splits, until it arrives at the gates (symbolized in Fig. 5.11 with the line width of the runners). This will ensure that all cavities are filled properly. However, the heavy runners are difficult to cool and add much to the plastic that must be recycled; this is wasteful, as explained earlier.

Figure 5.11, *right* shows a simple, schematic example for a similar mold in which the total runner system is divided into a (4-branch) hot runner system, and each branch will then supply a cold runner system, 2-plate in this example.

The pressure drop in the hot runner channels is small, and the final branches of the cold runner can be kept as small as they would be with the common runner layout shown on the left in Fig. 5.11.

(2) For very large products (Fig. 5.12) as found, for example, in the automotive industry, the product should be edge gated. The edge gates are located where best suited for the product, but instead of having a large cold runner supply, these edge gates, a (usually nonstandard) hot runner manifold, or other method of ducting the hot plastic will bring the plastic to the gates (or group of gates), without significant loss of pressure and without the need to reprocess the heavy runners. Another advantage is that the product can be placed approximately symmetrically around the center of the machine, for a balanced clamp force. This is possible with standard 3-plate molds, but not with 2-plate molds because the machine nozzle is in the center of the platens. (*Exception*: Special molding machines equipped with an offset extruder or one that can be located outside the platen and injected in the side of the mold or even into the P/L.)

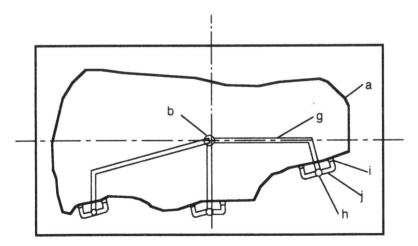

Figure 5.12　Schematic of a large molding, with 6 cold runner edge gates and a hot runner system with nozzles into each of the cold runners: (a) product, (b) sprue, (g) hot runner branch, (h) hot runner nozzle, (i) gate, (j) cold runner.

5.2.4.6   Insulated Runner Molds

Figure 5.13 (*left*) shows a mold similar to that in Fig. 5.9, but the sprue is an insulated runner. The plastic within the runner will stay hot long enough that the material injected during the next cycle will "shoot through" the still hot plastic, even pushing the, by then, frozen gate out of the way, but only if the cycle is not too long. Cycle times of up to 30 seconds can be successfully handled, and with some materials even longer, before the plastic freezes. If the plastic freezes, the cold "plug" is easily extracted after retracting the machine nozzle; as soon as the machine nozzle is again in position, the next cycle can be started. This method is simple, inexpensive, and reliable, and it is not sensitive to dirt. If some dirt blocks the gate, it can be easily removed, as if the gate were frozen. The gate can be very small but must be properly designed for shape and size.

    Figure 5.13 (*right*) shows a mold similar to the schematics in Fig. 5.10 (*right*). The hot runner is replaced with a much simpler insulated runner channel (e). In this system, the plastic in the center of the runner remains hot, even though the plastic close to the cooled walls will freeze; successive injected plastic will be able to flow through the hot core of the runner system without any added heat. It works well at cycle times up to 15 seconds, and even longer, depending on the plastic used. This system is very inexpensive, simple, and reliable, but the start-up procedure is somewhat awkward and possibly dangerous if performed by operators not skilled in this system; often several starts are needed before the mold will run on cycle. If the runner freezes, the mold must be split open between the cavity plate (a) and the backing plate (c); the, by now, frozen runner must be removed; and the plates locked together again before restarting. Molds with up to 16 cavities have been built and run successfully, but it is better to stay

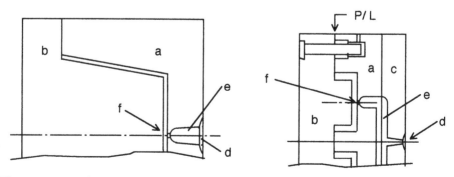

Figure 5.13   Schematics of single and multiple insulated runner molds: (a) cavity, (b) core, (c) backing plate, (d) nozzle seat, (e) insulated runner, (f) gate.

with not more than 6 cavities. Color changes are very easy. Without ever stopping the machine, by just changing to a new color in the extruder hopper, after about 15 shots, pieces with the new color are produced. More about this in *IMT*, p. 57.

5.2.4.7   Common Rules for Runner Systems

There are some basic rules to consider that apply to any and all runner systems. Unfortunately, some of these requirements are contradictory, and in such cases the best compromise must be found.

- **Rule 1:** *Pressure drop.* There should be a minimum pressure drop between the machine nozzle and the cavity space, after the gate. This affects selected runner lengths, runner cross sections, and gate size. The longer the runners, the smaller the runner cross sections, and the smaller the gates are, the higher will be the pressure drop; therefore, less pressure is available to fill the cavities. This means that often the length and thickness of runners must be increased, thereby increasing the inventory (see rule 2) in the case of hot runner molds. This will increase the time required for cooling the runners, in the case of cold runner systems, because of the larger mass of the runners.

- **Rule 2:** *Plastic inventory.* In hot runners, there should be as small a volume (inventory) of plastic as possible in the system between machine nozzle and cavity. The larger the runners (less pressure drop, see rule 1), the greater will be the inventory, and the longer the plastic will be exposed to the heat in the hot runner manifold, which can degrade the plastic within the runner system. The time for each temperature before the plastic starts to degrade is different for each plastic and is shown in graphs that can be obtained from plastics materials suppliers. Some plastics are very heat sensitive; others are not. For most plastics, it is desirable to have the inventory not larger than between 2 and 3 times the total mass of one shot, so that the plastic within the manifold is continually replaced, thereby reducing the length of exposure to heat. This is especially important with slow cycles where the plastic resides for a long time in the manifold; the same applies when molding very small products.

- **Rule 3:** *Heat loss.* There should be a minimum heat loss between machine nozzle and cavity. Heat loss affects the melt temperature and increases the viscosity of the plastic, making it harder to inject.

■ **Rule 4:**   *Cold runners only.* The area around sprue and all runners should be well cooled, for shortest molding cycles. In cold runner molds, poor cooling of the runners (little heat loss of the melt) means longer cycles, while waiting longer until the runners are cool (stiff) enough for ejection.

■ **Rule 5:**   *Hot runner molds only.* The hot runner system should be well heat insulated from the surrounding plates. Some heat losses are unavoidable because the hot runner manifold must be well supported (against injection pressure) by its backing plate, and by the features necessary to locate it within the mold. These necessary areas of contact conduct heat away from the hot runner to the surrounding cooled plates. Heat may have to be added through the hot runner manifold heaters to make up any heat losses, thus increasing the electric power used. Also, the heat loss into the surrounding plates can raise their temperature and affect the mold alignment; good cooling is necessary for these plates.

■ **Rule 6:**   *Balanced runners.* In any runner system, the pressure drop from the machine nozzle to each cavity space (gate) should be the same. (See also rule 10.) Pressure differences from cavity to cavity will affect the amount of plastic entering the gate before it freezes, the density of the plastic in the cavity space, and thereby the strength and quality of the molded piece. It will also result in differences in the surface definition, and appearance of the product. It is not always possible to follow rule 6 completely, but every effort should be made to do so. In some cases, individual adjustments to gate sizes may help to ensure more uniformity of the products.

■ **Rule 7:**   *Number of gates per cavity.* Wherever possible, there should be only one gate per cavity. There are exception to this rule: (1) where core shift could be a serious problem, two or more gates may be located symmetrically around a delicate core to equalize pressure and flow around the core; (2) where the flow length L from the gate to the farthest corner (or rim) of the molding is very great. This applies especially to large moldings. See also Section 5.2.4.8.

■ **Rule 8:**   *Location of gates.* Gate location depends on the shape of the product. A general rule is that the distance from the gate to the farthest corners of the cavity space should be about the same. Ideally, gating in the center of the product will fill this condition, but this is often more expensive than edge gating. In some cases, center gating is not acceptable if the center of the product must be clear or does not permit a gate vestige.

■ **Rule 9:**   *Breaking up the plastics flow.* Preferably, gates should be located so that the stream from the gate is broken up as soon as it enters the cavity space, by colliding either with an opposing wall or, at least, with some projection (such as a pin) in the cavity space. This will prevent the effects of jetting, that is, visible flow lines of the plastic, or other surface flaws.

■   **Rule 10:**   *Avoid reversed flow* (if possible). Gating into ribs or other heavy sections may cause the plastic to flow easily and quickly around some thinner areas of the cavity space. This creates additional fronts flowing toward the front of the stream coming from the gate; it will trap air, which must be vented. This occurs sometimes in heavier moldings, and venting can often be provided by judiciously placing ejector pins or vent pins at such locations where the plastics fronts are expected to meet.

5.2.4.8   L/t Ratio

An important characteristic of any product (and the cavity space) is the L/t ("L over t") ratio. This is the distance from the gate to the farthest corner (or rim) of the product, divided by the typical wall thickness through which the plastic must flow. It applies not only to cup-shaped products, such as containers, but also to flat products, whether they are center or edge gated. For example, a container, center gated in the bottom, has a distance of 300 mm from the gate to the rim. The wall thickness is 2.0 mm. In this case, the L/t ratio is 300 divided by 2, which equals 150 ($L/t = 150$).

From experience, it can be stated that an L/t smaller than 100 is usually easy to fill and (with some exceptions) a ratio of 100 to 200 is more difficult to fill. Any ratio above 200 is difficult to fill and may require special attention; it may even be impossible to fill. In the case of large products, by increasing the number of gates and spacing them judiciously, the L/t ratio can be reduced, and a piece can be produced that would otherwise be impossible to mold. Reducing the L/t ratio in this way would also allow the mold to cycle faster. Venting must be carefully considered to ensure that any air trapped between the fronts of the plastics flow from more than one gate will be able to escape.

5.2.5   Venting

5.2.5.1   What Is a Vent?

A vent is a small gap in the molding surface, located where air is expected to be trapped by the advancing and/or converging streams of plastic, during injection.

5.2.5.2   Design Rules That Apply to all Molds

■   **Rule 1:**   *How much venting?* Always provide as much venting as possible! The parting line is the ideal location for vents, but other areas are often as important. Especially with thin-walled, typically, disposable goods, but also with others products, the molding cycles depend on the speed with which air can escape the cavity space.

■   **Rule 2:**   *Ends of the melt flow.* For good filling of the product, provide vents where the melt flow is expected to end, and at any and all other points where air could be trapped, thereby preventing the melt to enter there, typically, at deep bosses or ribs. (If ribs end at the P/L, there is usually no need for additional vents.)

■   **Rule 3:**   *Weld lines.* Consider locating vents at points where two or more fronts of plastic flow will meet (for example, where weld lines are anticipated). This is important when using more than one gate into one cavity space, or where it is expected that the melt flow from one gate will split and then reunite, for example, when large cross sections in the cavity space cause the plastic to run around an enclosed area. Remember, all fluids always travel the path of least resistance.

■   **Rule 4:**   *Vent gap.* The gap must be large enough to let air pass, but small enough so that the inrushing plastic cannot follow. There are several considerations when designing the vent. Its size (the gap) will obviously depend on the viscosity and the pressure of the plastic. Commonly used are gaps of about 0.01 mm (0.0004 inch).

■   **Rule 5:**   *Land, vent grooves, and channels* (Fig. 5.14). The distance in the gap through which the air has to squeeze is called the land. It is good

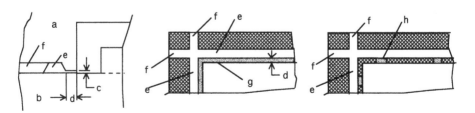

Figure 5.14   Examples of vents: (*left*) section through vent, vent groove, and vent channel, (*center*) continuous vent, (*right*) spot vents. (a) Cavity, (b) core, (c) gap, (d) land, (e) vent groove, (f) vent channel, (g) continuous vent, (h) spot vent.

practice to make the land very short. The length suggested for most molds is 1.5 to 2.0 mm (0.060 to 0.080 inch). (Longer land is, of course, possible but will offer more resistance to the escaping air.) However, the best-designed vent will not function if the air cannot go anywhere. As the air escapes through the vent gap, it must be permitted to flow away from the mold; the land should end in a vent groove, running approximately parallel to the edge of the product, and vent channels leading away from specific vents or from vent grooves. For venting at the P/L, the cross section of the vent grooves and channels should be commensurate with the amount of air expected to flow through them, at least 1 mm (0.040 inch) deep × 2 mm (0.080 inch) wide. For vents not at the P/L, the land should connect to a hole leading to the outside. This applies to fixed vent pins, and venting where two fixed mold components have a vent cut, for example, at the bottom of a deep rib.

■ **Rule 6:** *Width of gap.* There are spot vents and continuous vents. Spot vents were used commonly in earlier days of mold making. The molder noticed spots where the plastic was burnt at the edge of the product; where the burning occurred, a small vent at the P/L was cut into the mold, often crudely, with hand tools. Today, the mold designer must anticipate where spot vents will be required and specify their width. The vents can be as narrow as 2 mm (0.080 inch) or even less, but are more often about 6 mm (0.250 inch) wide. Continuous vents on the P/L are often specified for high-speed molds where they allow air to escape quicker than through a number of spot vents. It does not matter where the vents or channels are located on the P/L; they can be on the core side or the cavity side of the mold; the deciding factor is the ease of machining (grinding) them into the mold.

■ **Rule 7:** *Cleaning of vents.* Consider how vents are to be kept clean. Most plastics exude sticky substances that over time plug the vents. The vents in the P/L can be easily cleaned by wiping from time to time. Ejector pins and sleeves have clearances suitable for good venting and, because of their motion while ejecting, are considered self-cleaning. Specially designed vent pins are fixed in their locations and will have to be cleaned from time to time to ensure proper functioning. Frequently, the vent pins or other vents inside the mold are connected with drilled holes, not to the outside, but to a permanently pressurized air supply that blows through the vents when the mold is open. It does not affect the molding because the injection pressure is many times greater than the air pressure.

■   **Rule 8:**   *Strength of P/L*. The designer must not forget that the vents and the vent channels reduce the area where cavity and core meet (the P/L). The designer must make sure that in strength calculations referring to the compression of this area when clamping the mold, the actual area of the P/L is considered. This is sometimes overlooked, and after a few months of operation, due to fatigue of the mold steels, the cavity or core surfaces meeting at the P/L are compressed to such extent that the vent gap is reduced or even eliminated; a mold that ran fine at first gradually stops producing good products and will require recutting of the vents.

5.2.6   Ejection (See also *ME*, Chapter 12)

This is the next step in the design of the stack. As discussed in Section 4.9, there are many ways to eject a product. At this point in the design process, the designer will determine which method will be most suitable for (1) the shape of the product, (2) the type of mold, and (3) the expected productivity. The selected method will now be shown in proper relationship to the (cavity and core) stack. Space requirements for ejection mechanisms, including the location of the now also selected ejector plate return features, can influence the spacing of the stacks in a multicavity mold. The designer must also consider that the core must be backed up against excessive deflection of the core backing plate during injection. This backing up is usually simple with stripper rings or plates; almost the whole area under the core can be well supported because there are no ejector pins or sleeves there. It is often quite difficult to locate the ejector pins in the most effective locations, while allowing sufficient space for the backing up of the core plate and for aligning and guiding the ejector plate. Note that all ejector plates must be guided independently; that is, these plates must not be guided by ejector pins or return pins, because the weight of the plate will tend to bend these pins in the (usually horizontal) molding machines. But even in molds to be run in vertical machines it is good practice to guide the ejector plates.

If it is not possible (because of close spacing of ejector pins) to provide direct backing support, such as support pillars, under the core plate, the only solution is to provide very thick, heavy core backing plates to minimize deflection. The plate thickness can be calculated with complicated but accurate methods, or approximated as shown in *ME*, Chapter 17. The designer must also consider that after the mold cooling has been decided, it still may be necessary to relocate some ejector pins or some cooling channels. This may take several attempts of layouts before settling on a final solution.

5.2.7   Cooling

This section does not go into the details of mold cooling, but only highlights the most important areas and principles to be considered by the designer. For more information, the designer should consult *ME*, Chapter 13.

5.2.7.1   Purpose of Cooling a Mold

(1) *Cooling is directly related to productivity.* An injection mold could also work without any cooling; that is, it could rely entirely on giving up the heat energy, which was put into it during injection of the hot plastic, to the surrounding shop (ambient) temperature. This could take a very long time, especially with heavy sections and large masses of plastic, but it is done occasionally if the total production is very small. Instead of water cooling, air could be blown at the hot mold surfaces to cool them and to speed the process up somewhat. This is sometimes done even in production molds, when it is not possible to cool a very delicate mold part by conventional cooling means.

(2) *Productivity.* The higher the productivity that is expected from a mold, the faster the mold must be brought back to its optimal operating temperature, that is, the better must be the cooling. As should be stressed again and again, the molder is really interested only in getting the best product at the lowest cost, and the mold cost becomes significant only if production is fairly low. This means that, while a relatively low production mold should be well cooled, it should be done without "going all out"; with high production molds, there should be no limits to ingenuity when designing the cooling channel layout or selecting the mold materials for good conductivity and mold life. This is a typical area where compromises may be necessary. In certain types of molds (especially molds for intricate technical products, even for high production), the cooling of the mounting or backing plates is often acceptable, without any intricate cooling channels within cavities and cores. The heat must travel through cavity and core to the surface where they are mounted, and be removed by the plates by a pattern of simple (drilled) water channels. This method is inexpensive but will add time to the molding cycle, when compared to complicated cooling layouts. However, the added cost to the molding process is small compared with the possibly much higher mold cost with intricate cooling arrangements.

(3) *Heat conductivity of mold materials.* The designer must understand that there are great differences in heat conductivity in the materials commonly used for molds. The designer must also understand that the amount of heat removed

per unit of time depends on the distance the heat must travel. Dirt and corrosion in cooling lines also act like heat barriers and affect the heat transfer from the mold to the cooling medium. In some cases, it may be almost impossible to provide cooling lines in small mold parts; small pins and blades, for example, will heat up much more than the well-cooled cavity walls, and thereby control the molding cycle. Special mold materials, such as beryllium–copper (BeCu) alloys, provide about four times better heat conductivity than steel, and are often used for such delicate mold parts; the heat will then move faster than in steel to reach a well-cooled mold part or plate. Certain larger parts in high-speed molds are also made from BeCu, wherever it is important that heat be removed fast, even though such parts can be well cooled by cross drilling or are surrounded by coolant channels. This method is often used in mold parts opposite the gate where the hot plastic hits first as it comes out of the gate, and in parts surrounding the hot gate. Note that BeCu is much more expensive than mold steel; it can be used prehardened at about Rc 35–38, which is in many applications sufficient. BeCu, even when hard, is not as resistant to wear as hard mold steel; gates if made from BeCu must be replaced frequently, as the plastic stream tends to wash out and increase the gate size. The designer will make sure that such replacement is easy to do.

Caution: BeCu gives off poisonous gases during machining, and special precautions, such as ventilation of the work place, are necessary.

(4) *Heat conductivity of molding materials.* Plastics, too, have different heat conductivities. There is also the difference between crystalline and amorphous plastics. Crystalline plastics (e.g., PE or PP) contain more heat and give it up slower to the coolant than amorphous plastics (e.g., PS); without going into details, more energy (heat) is needed to melt crystalline plastics, and more energy (in cooling) is needed to cool it down again. In practical terms, for example, by using the same mold, it will take longer to mold a product from PE than from PS.

As soon as the plastic touches the walls of the cooled cavity space, it freezes, which makes it more difficult for the following layers to give up heat to the mold wall or insert. This is significant for products with heavy walls and will increase the cooling time regardless of how well the wall (or mold part) is cooled. Also, as soon as the plastic begins to cool, it will start to shrink; this happens (in most cases) in the direction away from the cavity. Because the shrinking plastic will start to hug the core, there will be (a) a better contact with the core, and (b) a space created between the plastic and the cavity wall; this space contains a vacuum or if properly vented, will contain air. Both air or vacuum are ideal heat insulators and reduce the heat flow from the plastic to the cavity wall. In most such molds the cavity does not need as much cooling as the core. Unfortunately for the designer, there is a problem: usually, there is much more space in the cavity—where cooling is not needed as much—to provide lots of cooling circuitry, while the core—which does more of the cooling—is much more difficult to cool, especially when there are also ejectors, moving parts, and/or air channels in it. Many existing molds have lots of unnecessary cooling in the cavity.

With molds for thin-walled products, it is somewhat different. The injected plastic is so thin that there is less effect of shrinkage, and the cooling of the cavity also becomes important because the plastic stays in contact with the cavity walls for much longer.

There are exceptions to the foregoing. For example, if a product has heavy walls and a large gate, injection pressure can be maintained longer, the shrinking volume is replenished during the cooling cycle, and the plastic stays in contact with the cavity wall longer. Even so, the cavity cooling is never as important as the core cooling.

5.2.7.2   Show Cooling Lines in Stack

The next step in the design is to show the selected cooling lines in the stack, that is, the cavity, core, and, occasionally, the stripper plate and any side cores or cavity splits. This may require several attempts of layouts before settling on one solution. For very high production molds, this may take considerable design time but it is always worth it. It may also require going back to the stack layout and changing the ejection layout to arrive at a good compromise in locating both ejection and cooling. As mentioned earlier, make sure that all channels are dimensioned so that the coolant will have turbulent flow and that the location of channels from the molding surfaces is as suggested for efficiency and strength.

How will the coolant be supplied to the cavities and cores, in case of multicavity molds? There are several possibilities.

(1) Each cavity or core is mounted on its respective backing plate, and each has its own coolant connection to a central water supply (header, etc.) This is fairly inexpensive, but not very good because of the large number of hose connections required, especially when there are more than six cavities. Remember that every cooling circuit has an IN and an OUT connection (2 hoses), and often there are several cooling circuits per cavity, and, similarly, often more than one cooling circuit per core or cavity. In addition, some plates should also be cooled because of possible alignment problems. All this can add

up to a very large number of hose connections, a possible nightmare for mold installation.

(2) Cavities and cores receive the coolant from their underlying plate. This method is more complicated than (1), but reduces the number of hoses required to a minimum. The mold plates are cross drilled with channels of various (larger) sizes to supply the coolant and to return it. These sizes should be calculated and located so that all cavities or cores will be able to draw, as nearly as possible, the same amount of coolant. Cross-drilled channels are more expensive to produce than the method shown in (1), but such molds are much less troublesome to install, or in operation. Note that the coolant should not be used to regulate the flow through some portions of the mold during the operation of a mold. The coolant should be either ON or OFF. In exceptional cases, it may be necessary to shut off the cooling around hot runner nozzles during start-up, but even this is old-fashioned and unnecessary if the mold is properly designed.

(3) The cavities are often inserted (fully or partly) and therefore fixed in position. The cores are usually screwed on backing plates, sometimes even allowed to float, for perfect, individual alignment. For the coolant connection, the same applies as in (2).

Regardless of which of the above three methods are used, the designer must now consider where the coolant connections are located in those stack members that will be cooled. It is very desirable (for mold making and for servicing) that all stack parts are the same; the designer should spend some time to see if all parts can be mounted without the need for "right" or "left" parts. In cases (2) and (3), this can often be achieved by judiciously locating the coolant channels. To prevent leaking, O-rings will be required at all fluid passages from one mold part to another. O-ring grooves and finishes must be properly specified. In some cases, more than one passage may be covered by one O-ring (use O-ring manufacturers guidelines). Any leakage from one passage to another within the O-ring ("wet") area can be ignored, but it is important that no screws are allowed in a wet area.

5.2.7.3   Screws

At this time only, the designer will locate the screws connecting the cavities and cores to their plates. In some cases, where cavities or cores are inserted in plates, they can be held in them with "heels," and, therefore, do not require screws; but if the inserts are round, they must be oriented, for example, with dowels, so that they cannot turn. If screws are used they too should be located so that there is no need for "right" and "left" parts. The designer should always make sure to use the lowest number of screws required to contain the expected forces that the screws are supposed to withstand, and to select the largest screws possible in that location. In manufacturing, as a general rule, any screw thread smaller than 8 mm diameter (while of course possible) is more costly to produce. From experience, most molds have too many, often unnecessary, screws. Note that the foregoing applies for all screws in the mold, not necessarily the stack. (See also *ME*, Chapter 19.)

5.2.8   Alignment of Stack

This should also be decided now, before proceeding. Will the overall alignment of the mold shoe with leader pins be enough? Should each stack be aligned by taper locks? By a pair of leader pins? For this decision, see Section 4.11.

5.2.9   Design Review

This is a good time to sit back and contemplate what has been achieved so far. Is it really the best thing the designer could come up with? Please note that all the things discussed up to now in this text are, or at least should be, in the head of the experienced designer, and all the work done up to now would normally not take more than a few hours for an easy mold or maybe a few days for a more complicated one. This is also the time that the designer arranges for a design review, as discussed earlier. The result of such review will then determine whether to proceed as shown, or to "go back to the drawing board." Often, only minor changes may be required, but frequently, as experience has shown, new ideas come out of these meetings, and the result will be a better operating, and maybe a lower cost, mold.

5.3   Preload

The term "preload" has been mentioned several times in our discussion. What is preload? As an example, imagine two blocks that are held together by two screws. These blocks are subjected to a force F trying to separate them. If both screws are hand tightened, that is, tightened just enough that the blocks touch, without any gap between them, the screws will not exert any force SF on the blocks; the combined total screw force *SF* equals zero ($SF = 0$). As soon as the force *F* is applied, and because *F* is greater than *FS* ($F > FS$), the blocks will separate and the screws will be stretched until the resistance (or force) in the screws *SF* equals *F* ($SF = F$). But by then, the blocks have separated and left a gap between them. In molds, any undesired gap means flashing or leaking, and is not acceptable. To prevent such gaps, the screws must be tightened to such an extent that they will be stretched to a desired preload. *FS* must be greater than the expected force *F* ($FS > F$). When the force *F* is now applied, the blocks will not separate unless *F* becomes greater than *FS*. In practice, there are two types of preload.

(1) *The preload exerted by screws.* Screws must always be tightened to the manufacturers suggested values, that is, to about 60–70% of the yield strength of the screw. The resulting force (or holding power) of the screw can be found in all screw tables.

(2) *The preload can be provided by stretching the steel of mold parts*, such as tapers, wedges, stripper rings, and so on, or *mold plates*, as in the following example, and by press fits, which are a kind of preload, or by shrinking of rings or bars over—usually—cavities, for building up cavities from sections. When specifying preload on tapers or wedges, it is common practice to indicate the distance (which, unfortunately, is also called preload) that the tapers are allowed to move (and thereby stretching the steel) before coming to a stop. This preload is especially important where cavities split in two or more sections.

For example, a mold for a mug with handle (see Chapter 7) will split in a vertical plane through the handle. If the two cavity halves are not preloaded, the splits will open under the injection pressure and the mold will flash both at the handle side and at the side opposite the handle. In this case, the preload is provided by having the cavity sections backed up by wedges, preferably both on the cavity and core side, which will make contact with the cavity sections before the mold is fully closed. As the mold closes fully (over the length of the "other," calculated preload) the wedges stretch the cavity plate and (preferably) also the core plate. The stretching of these plates provides the necessary "real" preload (in kN or US tons) to hold the mold together against flashing.

Preload is explained in much detail in *ME*, Chapter 30.

5.4   Mold Materials Selection

At this time (or maybe even earlier, while designing the stack), the designer will think of the materials (steels, etc.) to be used for the mold. (See also Chapter 9)

5.4.1   Effect of Expected Production

Before making any decision, the designer must again consider the lifetime production expected from the mold. There is no point in specifying the best possible (and expensive) materials if the mold will be required for a small production. Also, there is a difference whether, for example, 24 million pieces are to be produced in a 24- or an 8-cavity mold. With 8 cavities, the mold will operate 3 million cycles; with 24 cavities, it will operate only 1 million cycles. This requires the designer to consider fatigue in metals, as discussed in Section 5.4.3.3.

5.4.2   Forces in Molds

The designer must know what forces are present within the mold when deciding on the strength of the mold component to resist these forces. The most important forces acting within the mold affect these strengths:

(1)   *Tension:* the forces created by the injection pressure of the plastic inside the runner system and in the cavity space, usually requiring high tensile strength
(2)   *Compression:* the compressive strength required to counteract the clamp force of the machine, typically, the forces on the P/L, and the forces seen where inserts are supported by plates, and so on
(3)   *Bending* (or deflection): the forces seen by cores, and by all plates, especially the ejector and stripper plates
(4)   *Wear:* the forces created by wedge action, as in stripper rings and so on, or tapers and wedges for alignment, which create wear on the matching surfaces
(5)   *Torsion:* the forces seen by coil springs and in mold features, such as unscrewing, or in some robots
(6)   *Shear:* forces seen by dowels, or by the backup of wedges

Note that in many cases, we have combinations of any of the above forces.

5.4.3   Characteristics of Steels and Other Mold Materials

For mold steel selection, see Section 9.2.

For every mold part the following must be considered: which of these characteristics are most important? Unfortunately, some of them are directly opposite to each other (e.g., toughness and hardness) and compromises are necessary.

5.4.3.1   Availability

This applies not only to selected raw materials, but also to hardware items: the designer must make sure that any material, hardware, or standard mold component intended to be specified is also available when required. Many items are often shown in catalogues or other listings as "standard" but this does not always mean that they are readily available, on the shelf, in the desired size, and in the quantities needed.

5.4.3.2   Strength of Material

This applies to steel, BeCu, aluminum, bronze, and so on. Strength is specified by its tensile strength; compressive strength is often but not always about the same. Shear and torsional strength is about one-half the tensile strength. The designer should always get the exact values from a machinery handbook or from the supplier.

Always watch whether the values given are in ISO or in inch systems. The strength values are given either in kPa (kilopascal) or in psi (pound/in^2).

5.4.3.3   Fatigue (See *ME*, Chapter 18)

The strength figures for steel and other metals are arrived at from stressing a test sample, for one cycle only. The results of such tests are satisfactory for steady loads, such as seen, for example, by preloaded screws, but molds often operate many, sometimes millions of cycles. If there are more cycles, the rated strength gradually declines.

This decline is usually shown, as in Fig. 5.15, in logarithmic graphs, as a straight line declining from the rated strength (e.g., tensile or yield strength) for one cycle to a point where the value remains the same regardless of the additional number of cycles; this is for all steels at about 2 million cycles. The

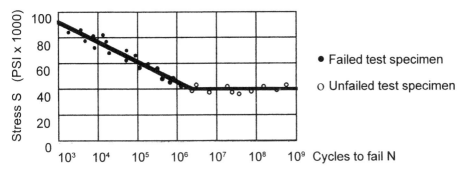

Figure 5.15   Typical fatigue graph for a machinery steel.

strength of the material, after 2 million cycles (the fatigue strength) depends very much on the material and hardness selected, but also on features such as notches, holes drilled into it, and surface finish. The fatigue strength can be as low as 15–20% of the yield strength (yield, in hardened mold steels, is only a little less than the tensile strength; many data are given in yield rather than tensile strength). Note that so-called machinery steels, but also the related P20 or P20PQ, do not lose as much strength as hard mold steels.

The fatigue strength is equivalent to the safety factor often used by designers (frequently, 5) when calculating the strength of a part. The problem is that all force calculations depend on an assumption of the injection pressure, as discussed in Section 4.6.1. But we know that the forces will be greater for thin-wall molding, and since most of them are designed for a very large number of cycles, the selection of only the very best materials with appropriate strength and hardness is suggested.

Note that springs inside molds (sometimes specified for ejector plate return) are especially sensitive to cycling. When designing for springs, use the manufacturer's suggested values for maximum compression and load of the selected spring.

5.4.3.4   Wear

Some materials are better for wear than others. Lubrication (or the lack of it) can be a decisive factor. Wear points could be steel on steel, steel on bronze, steel on hard plastic, and so on. Hard steels are always better, but the designer must never use the same alloy for both members rubbing against each other, as in wedges or taper locks, except if the wear points can be lubricated. Each alloy has a distinct, different grain structure, and the problem is that when using identical grain structures, the surfaces will lock (seize) when sliding under pressure, and damage (tear) the surfaces. Hardness differences alone are no substitute for different grain structure, except where one of the rubbing surfaces is treated with methods such as nitriding. In nitriding, very hard nitrogen compounds enter between the grains and alter the surface of the steel. Lubrication in molds is never permitted where it could contaminate the molded products, especially for pharmaceutical and food use.

5.5   Stack Molds (See also *ME*, Chapter 15)

All that has been said so far applies to any mold, single-level (conventional) or multilevel (stack). In principle, a stack mold is an arrangement where a number of single-level molds are placed back to back in the molding machine. Here, only the most common, two-level stack mold is discussed, although 4 levels and more have been built. The two injection (usually cavity) halves are mounted back to back in one moving ("floating") platen between the standard machine platens; the core halves are then mounted one each on the stationary and moving platens. (Because these are usually also the sides where the ejectors are located, special provisions must be made for ejector actuation on the stationary mold side; this is sometimes built into the mold.)

The stack mold system is often used for very large production, requiring many cavities, but often also for molds producing different parts that are paired in assembly. Stacks for one product are in one level, and stacks for another matching product are in the other level. The mold cost is about the same (or even a little less) than the cost of two molds, each built for half the number of cavities.

The *advantage* is that one stack mold on one machine, requiring much less plant space and investment, can have the same output as two molds, requiring two machines, provided that the clamp has sufficient stroke and shut height to separate both P/Ls far enough for ejection from both sides. Also, the injection unit must have a large enough plasticizing and shot capacity to fill both sides without increasing the cycle time, which, of course, would defeat the purpose of this system. Because the molds are stacked on top of each other, only the projected area of one level need be considered. The forces due to injection pressure within the center plate cancel each other; however, it is suggested to use a machine that has a clamping force of about 10% more than would be required for an equivalent single-stack mold. Today, in most systems, the injection unit is connected with a long sprue extension to the hot runner in the center platen with the cavities. In some cases, the plastic is injected from the side, with a special extruder arrangement.

A *disadvantage* of the stack mold system is that in case of mold or machine trouble, with stack molds, there is no production at all, whereas with conventional molds, half the production will continue.

5.6   Mold Layout and Assembly Drawings

Now the designer has all the basic information about the mold to be built and can start to finalize the mold assembly drawing.

5.6.1   Machine Platen Layout

The platen layout—including tie bar locations—of the machine (or machines) the mold will be used on should be shown first. This will determine the outer limits of the mold and where to place certain mold features. It will, for example, specify where coolant connections must not be located, or any planned auxiliary actuators outside the mold, latches, and so on. The mounting and ejector holes that will probably be used for the mold must also be shown.

5.6.2   Symmetry of Layout, Balancing of Clamp

For multicavity molds, it is important that the stacks are positioned such that the projected area of each cavity is as symmetrical as possible about the center of the machine, to ensure that all tie bars are loaded equally as the mold is clamped, thereby providing each cavity with the same preload to prevent flashing. This can present a problem with "family molds," where several different stacks or cavities with different projected areas are used in one mold. A small amount of asymmetry is often acceptable. With edge-gated, single-cavity molds, to balance the load, a pressure pad must be used opposite the stack location to simulate the force of a second cavity. In this case, the cavity itself will see only one-half of the clamping force of the machine. This is important for the selection of size of clamp, for the job. There is no such problem with center-gated, single-cavity molds.

5.6.3   The Views

Start with the significant mold cross section or sections, but always work with all views at the same time; that is, both the plan views of cavity and core will "grow" side by side with the cross section. This prevents surprises arising when one view is far advanced and then it becomes apparent that it does not go together because another view shows some interferences. Show the selected hot runner hardware, if this is planned to be a hot runner mold. If it is a mold for which the hot runner section is purchased completely assembled by the supplier, show the interface points and dimensions only.

5.6.4   Completing the Assembly Drawing

Everything can now be shown in all views. It is not a good practice to show the complete stack in every location, even though it is easy to do with a CAD system. It would make it difficult to read the drawings, especially if there are many other features in the stack. To facilitate the reading of the drawing, the stack should be shown in only one location of each plan view, and just its outlines in all other locations, for example, with heavy, dotted lines. However, important information such as the centers of coolant connections, screws, alignment features, and so on should be identified in all locations with small crosses and/or circles, which can then also be identified with a code, such as S1, S2 for screws and D1, D2 for dowels. Such codes will make it easier to read the drawing; they will be also important when completing the cooling lines layout in the plates and the location of plate supports and large screws holding together the various mold plates, where applicable. Show now also the alignment features, the ejection system, the method of mold mounting and any connection (fixed or loose) with machine ejectors, and everything else needed by the detailers to produce the shop (detail) drawings.

At this time, show also where the outside of the mold must be marked (preferably die stamped) to identify coolant and air connections. There would be a 1 IN, 1 OUT, 2 IN, 2 OUT, and so on, and AIR 1, AIR 2, and so on. The IN and OUT can be important for cooling because in many cases it does make a difference where the cold coolant should go first (IN). For example, in the core of a container mold, it should first hit the area opposite the gate.

5.6.5   Bill of Materials (BoM) and "Ballooning"

This is also the time to specify the BoM so that all materials can now be ordered and be available when required for the machining operations and the final assembly. The BoM should specify not only the final sizes of steels and so on, but also the hardness of the finished mold part. This is important not only for the buyer, but also for the detailer of the shop drawings.

"Ballooning" is the identification of each mold part on the assembly drawing. Several methods are used, but the preferred one is to show balloons (circles or ellipses about 12–15 mm in size) outside around the drawings. Each balloon contains a number identifying each mold component, but only once, from stack parts to plates to screws, and so on. This number corresponds to a line in the BoM. Each balloon has a leader (line) connecting it with the part identified. Preferably, the balloons should be shown around the main cross section of the mold or near partial sections; only if these locations would not be clear enough and could cause errors should they be shown in other sections or in the appropriate plan view.

5.6.6   Finishing Touches

Finishing information of the molding surfaces should also be shown— preferably with standard symbols—on the assembly drawing, for future reference, and to be used by the detailer when making the shop drawings. Cross hatching should be used sparingly, only where it really helps to make the assembly drawing clearer. This also applies to detail drawings. This is also the time to show any notes on the drawing. (See also Section 5.2.2.4)

Usually one "main" title block is shown, preferably on the drawing with the main cross section; additional, smaller title blocks are on all other drawings. The title blocks identify the mold design office or the mold maker, the project number and drawing numbers, the designer (by name and initials), the checker, and the detailer, if applicable. It will also show any other information pertinent to the product and will specify for which machines the mold was designed, the types of plastic, and any other information that deserves to be recorded for future use. Tolerances are not shown on the assembly drawings. They are strictly limited to the detail drawings. However, it is a good practice to show fits and clearances where they apply, but only if they are different from standard fits and clearances.

6 Review and Follow-Up

After the drawings and the final BoM have been released for production, there is usually a quiet time for the designer, as far as this mold is concerned. Hopefully, there are no problems in buying and machining. If there are problems in the shop, for example errors in machining or—heaven forbid—errors in the drawings, any corrective action must be approved and recorded by the designer or his delegate. There is always the possibility that the same mold will be required again maybe in a year, or much later, and it would be embarrassing if the same errors would then be repeated. After the mold is finally ready for testing, the designer must be present and see that the installation and setup procedures are in accordance with the specifications on the assembly drawings. The designer must also approve any changes required to make the mold work as expected and record what was done to make the mold work before it is shipped. A complete report, specifying the test machine, all temperatures, times, pressure settings, and plastics specifications should be supplied to the customer, together with the mold.

A good designer will then follow up the mold with the molder, especially in case the designer has not heard from the molder first, to see how the mold works in production. Unfortunately, frequently, a mold goes into the customer's molding shop, and if there is any problem, the shop people cannot be bothered to go back to the mold maker but make adjustments that may not have been necessary if they had followed the instructions received with the mold. Any later problems experienced by the customer should also be recorded for future reference.

7   Typical Examples

A few examples are provided of typical molded products and how they should be approached. These examples are used to illustrate material discussed earlier in the text.

7.1   Containers or Other Cup-Shaped Products

Containers are not necessarily drinking cups, but any container, round or of any other shape, such as boxes or many technical housings. The main characteristics of container molds are as follows: (1) Although they can be edge gated, they are usually outside center gated; they may have more than one gate. (2) Core cooling is usually easily accomplished, which is the basis of higher productivity. There are all kinds of shapes, too many to show in one book, but there are some significant typical differences. Some examples are shown here.

Figure 7.1 depicts two very similar cups: on the left is a typical cup (or container) with a plain bottom, and on the right is a cup with a reentrant bottom. Note that the bottom is preferably domed, as shown. While shrinking, the curvature of the dome will change somewhat but it will not pull inward and thereby deform the side wall of the container. It is always quite difficult to mold any straight surface, especially from high-shrinkage plastics, unless the cooling cycle is greatly extended to permit the product to reach the mold temperature before ejection. A typical mold for such a product is illustrated in Fig. 7.2. The gating can be a hot runner, 3-plate, insulated runner or through shooting.

Note that Fig. 7.2 shows a conventional mounting plate (17). As discussed in Section 5.1.6.1 (shut height), this illustrates a typical example where this plate can easily be omitted. The mold on the left in Fig. 7.3 uses a stripper plate, and the ejector plate comes to a stop when the stripper taper seats on the core taper, so the ejector plate does not need a stop. In the case of an ejector plate using

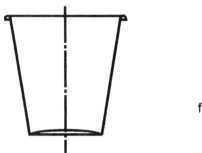 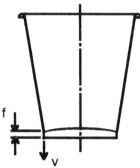

Figure 7.1 Schematic illustration of two typical cups: (*left*) a simple cup shape; (*right*) a similar cup but with a reentrant bottom.

ejector pins (right illustration), solid stops (shoulder bolts, etc.) must be provided; they can be mounted on the underside of the core backing plate.

In Fig. 7.3, the parallels and the supports under the cores (supporting pillars) will sit directly on the machine platen. The designer must make sure that when the mold is mounted in the machine, all pillars are fully supported; that is, they must sit on the machine platen but should not sit solely on top of any weak areas of the platen such as T-slots.

Note that in any mold, all the outside edges of mold plates, or any other area where sharp edges could cause personal injury during handling, should be properly broken (rounded or chamfered). However, in some areas, especially in the path of plastic flow, especially on inserts, sharp corners must be kept sharp; the designer must indicate this on the drawings.

The right illustration in Fig. 7.1 shows a typical cup with a reentrant bottom. Here, too, the bottom is preferably domed, as shown. But because of the reentrant, especially if the depth of the dimension f is greater than twice the thickness of the plastic at that spot, it will be difficult or even impossible to fill this portion of the bottom; also, if a piece of plastic breaks off in that narrow section and remains there, it would be very difficult to remove it without dismantling the mold. Therefore, special measures must be provided in the mold: the cavity of the mold must follow the core as the mold opens, for a short distance (about for the distance f) until the mold part that forms the inside of the reentrant, which usually also contains the gate, is completely withdrawn from the molded plastic piece. Only after this happens is the mold allowed to separate at the regular parting line. This method also facilitates good venting at the bottom, as indicated; otherwise, the thin section would be a "dead pocket" and not fill, as already discussed Section 5.2.5.2, rule 2. Note that this method is

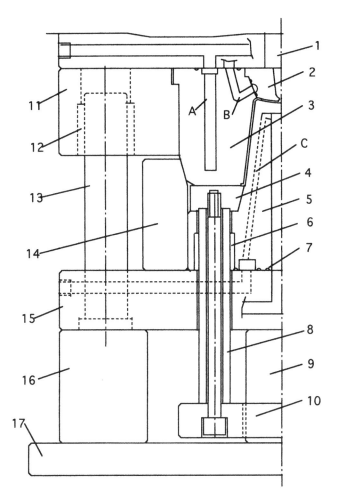

Figure 7.2   Schematic illustration of a section through portion of a simple cup mold: 1, back plate or hot runner plate; 2, gate pad with cooling; 3, cavity; 4, stripper ring; 5, core; 6, guide bushing for ejector sleeve; 7, O-rings; 8, ejector sleeve; 9, support under core; 10, ejector plate; 11, cavity retainer plate; 12, leader pin bushing; 13, leader pin; 14, locking ring (for alignment of cavity and core); 15, core backing plate; 16, parallel; 17, mounting plate; A, cavity cooling; B, gate pad cooling; C, core cooling.

called moving cavity (Fig. 7.4); it is, in principle, similar to the two-stage ejection illustrated in Section 5.2.3.3.

The cavity plate is guided on a separate set of guide pins to control its location relative to the gate retainer plate (or hot runner plate or cavity backing plate, as should be the case). Its stroke is limited to be only slightly larger than

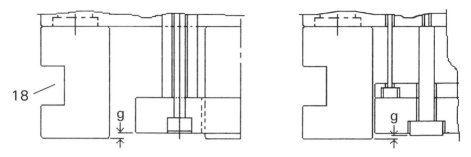

Figure 7.3 The elimination of the mounting plate of the mold assembly. Mounting slots 18 have been added to permit the use of mounting clamps. (*Left*) A variation to Fig. 7.2. (*Right*) This application for a mold with ejector pins. There must be always a clearance (g) where shown.

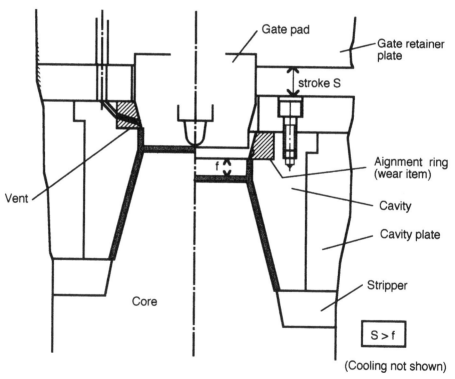

Figure 7.4 Typical construction of a moving cavity feature to release deep reentrants in the cavity. The left half shows the mold in the closed position, whereas the right half shows the mold at the point of opening when the cavity stops; the core continues to open until the mold is fully open. The product is ejected as soon as the cavity is sufficiently distant from the cavity half. Note the venting arrangement.

dimension *f*. Air actuators (usually four) built right into the backing plate push the cavity plate so that it follows the mold opening motion until the set limit is reached. The product is now easily ejected from the core, and there is no danger that the "foot" gets trapped between the gate pad and the cavity. There must be ample venting provided where the alignment ring meets the gate pad.

7.2 Technical Products

When designing molds for technical products, consider first: (1) gating and runners, (2) core cooling, and (3) alignment of cavities and cores.

(1) As discussed earlier, 2-plate molds with edge (or tunnel) gating are simpler and much less complicated and expensive than 3-plate molds or hot runner molds. They can be, and still are today, used in the majority of all molds, especially if the production is fairly low. The problem with edge gating is that any runner, leading from the sprue to the final branch runner (with the gates), must never be located so that it will have to cross an open space. This makes it necessary that all cavities and cores must be inserted in the cavity and/or core plate, with a perfectly smooth (but not necessarily flat) surface—the parting line—between them, without any gap into which plastic could flow. This also applies to any stripper plate with inserted stripper rings. Such rings, even though of great advantage for better alignment with the cores and ease of replacement, must not float in the stripper plate because of the obvious gap between ring and plate, a gap over which the runner would have to pass. The designer must decide whether to make rectangular or round pockets (or cutouts) into the plates, and (a) insert the complete cavities or cores with tight fit into them, or (b) cut the cavities (or even the cores) right into the plates and just place inserts, if required, into them. A round pocket will contain just one cavity or core; in a rectangular pocket, one or more can be packed (see Fig. 7.6). Many molds, from 2-cavity to multicavity molds, are built this way. This decision will also affect the choice of materials for the plates. Mild steels would be acceptable in one case (a) but usually not in the other (b).

The alternative is to gate into the top (outside) of the product, from the cavity, as with 3-plate, insulated or hot runner molds, where the runners are not in the parting line. With this choice, the cavities are frequently inserted into the cavity (or cavity retainer) plate or as individual units. The cores are usually individual units mounted on top of a core backing plate with gaps between them.

(2) The core cooling for technical products is usually not as simple as for containers, because of the often large number of inserts within the core or cavity.

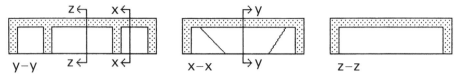

Figure 7.5 Schematic of a technical product, with inside ribs. One rib is as shown in section *x–x*, the other as in section *z–z*.

There is most often only one choice: to forget about intensive cooling with channels right into the cores or the inserts, and to depend on the heat conducted from the hot plastic, through the inserts and core or cavity, to the supporting, cooled plates (see Fig. 7.6). In some cases, better conducting materials, such as beryllium–copper, are used to make inserts or even complete cores or cavities. *Note*: Every gap (clearance), but even every area of changeover from one part to another, even when fitting tightly and without any gap, constitutes a heat barrier and slows down the heat flow. For this reason, most molds for technical products will cycle slower than the well-cooled molds for containers of similar weight and wall thickness.

(3) Multicavity, 2-plate molds with inserted cavities and cores (or where they are cut right into the plates) require high accuracy in the location of cavity and core, because there is no possibility of adjusting their relative position once the mold is finished. There is also the problem of heat expansion of the plates, which can shift the relative positions if the plates are not of the same temperature. For this reason, this type of mold should not be selected for thin-wall products where the wall thickness can be greatly affected by any misalignment. If high accuracy is required, it is best to have the cavities fixed in the cavity plate, and the cores mounted floating on the core backing plate, with individual method of alignment either with tapers as shown for a container, or, as is most commonly done, with additional, small leader pins and bushings in each stack. This will, of course, make it impossible to use runners in the P/L, and will require a mold with gating into the top of the product, as shown in (1) above.

A typical, technical product is shown in Fig. 7.5.

7.3   Mold with Fixed Cores

If a rib ends in a side wall as in section *z–z* (Fig. 7.5), venting of such rib is no problem since the sidewall ends at the well-vented parting line. If, however, the

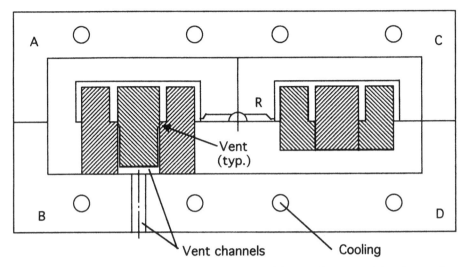

Figure 7.6   A schematic of an edge-gated mold, with two of more cavities shown. One cavity (*right*) has ribs as shown in Fig. 7.5, section *x–x*, the other (*left*) has ribs as shown in section *z–z*.

rib is "closed" as shown in section *x–x*, venting becomes very important, especially if the rib is "thin," that is, if the ratio of depth over thickness is greater than about 2–3.

The illustration in Fig. 7.6 could be a section through a 4-cavity mold. Both cavities A and C and cores B and D are set into pockets in the mold plates. Inserts (cross hatched) are located either in cutouts (core, left side), which is better for cooling, or in pockets (core, right side). Note, in the left portion of the illustration, that the venting channels for those ribs do not end in the side wall of the product. Note also that the runners sit on top of the line where two mold parts meet; they will not leak. Both cavities and cores are cooled from their underlying plates, as indicated by the circles, representing drilled holes for cooling. Note that the inserts in the left core are better cooled because there are fewer heat barriers.

7.4   Mold with Floating Cores

Figure 7.7 shows portion of a mold for a product similar to that in Fig. 7.5, but the requirements for accuracy are high, so the cores are mounted floating on the

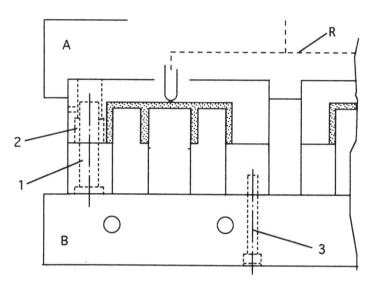

Figure 7.7 Schematic of a mold portion with floating cores. (A) Cavity plate with runner system (R) indicated with broken line. (B) Core backing plate. 1, Leader pin; 2, bushing; F, floating core mounting.

core backing plate (see *ME*, Section 14.4.2). The leader pins (1)—usually 2 per stack—are shown here with a bushing (2) in the cavity, but the bushing is often omitted, since the cavity itself is usually made from hardened steel.

Note that in these applications, with or without floating cores, the cavity is usually easier to cool, by cross drilling, than the core; however, as mentioned earlier in this book, there is not much gained by it because the core cooling usually controls the molding cycle. Much more can be gained by carefully considering where to gate, and providing ample venting in any area of the stack where air could be trapped.

7.5   Molds with Side Cores or Splits

For all molds with side cores or where the cavity splits into two or more sections, these sections must be preloaded against the forces from the injection pressure to prevent flashing along the split lines. Refer to Fig. 7.8. As the mold opens, the cavity "splits" move for a short distance with the core, while the splits open sideways. Only then can the cup be ejected. With the closed mold,

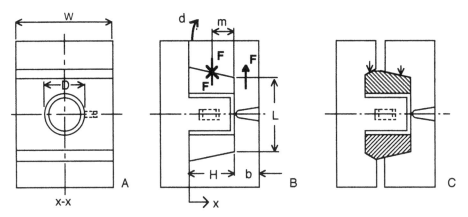

Figure 7.8   Schematics of a mold for a cup with handle: (A) plan view into the cavity, (B) section through a mold with wedges on the cavity half only, (C) a similar mold, but with wedges on both cavity and core sides. W, width of the plates; L, length of stretched cavity plate; b, thickness of cavity plate along L; H, height of cup; D, cup diameter; F, the forces to be contained.

during injection, the injection pressure p inside the cavity acts on the projected area of the sides of the cup, $F = p \times D \times H$. In mold B, the force F pushes against the wedge, which is part of the cavity plate and is counteracted by the steel of the cavity plate, with a cross section of $b \times W$. There are now two problems to consider: (1) the force F will stretch the portion of the plate with a length L, and create an undesired gap at the split line. The wedges must therefore be preloaded as explained in Section 5.3 of this book. (2) Because of the distance m between the forces and reaction forces, there will be a bending moment $m \times F$ which will force the wedge to bend outward as indicated (arrow d). This system is therefore only suitable for shallow products. For deep products, the side forces must be taken up on both the cavity and core sides of the mold. This is illustrated by mold C, which has wedges both in the cavity and the core side. The forces F trying to push the halves apart are thereby divided, and both cavity and core plates will provide reaction forces. The preload must be calculated and provided for each set of wedges.

8   Estimating Mold Cost

One of the most difficult jobs in the mold making business is to determine as accurately as possible the cost of the mold for the product for which it is to be built. The estimator should be an experienced mold designer who can visualize from the product drawing submitted (and occasionally from a sample) what kind of mold will be most suitable to produce the product economically.

8.1   Need for Estimate

Before estimating, the designer (and the person negotiating with the client for an order) should first establish if the "request for quotation," that is, to quote a price for such a mold, is serious and how the outlook is for getting the order. This is an important consideration: in the author's experience, many molders are often approached by their customers solely to find out how much it would cost, approximately, to start a new product line; they need a mold price to determine their own costs before proceeding. In some cases, the customer approaches not only one, but possibly three or more molders for mold prices, and each of these molders may in turn approach three or more mold makers for estimates of the necessary molds. One mold maker may then get the same inquiry from several molders, for the same product. In fact, only one of all these requests for estimates can result in an order. This means that the estimator, faced with all these requests, cannot spend too much time with each one, or the cost of estimating would become excessive. In many cases, the "boss" of the mold shop will decide whether it is really necessary to quote at all, or he or she may decide to just give a ballpark figure and skip the formal estimating process altogether. From the author's experience, with such multiple requests, the lowest price is often based on errors in quoting; with clients who habitually select the lowest bidder, the mold maker is bound to lose money. Any smart buyer of molds, before placing an order, should consider first the background and reputation of the mold maker and his or her expertise in building the particular type of mold requested. Only then should the price be considered. As has been said here repeatedly, only the best-suited mold for the planned production will result in the lowest product cost, which is really what the client needs. This often leads to specialization by the mold maker, which is beneficial to both customers and mold makers. Requests for molds that are outside the mold maker's expertise should be declined, unless the mold maker intends to enter this new field. If the request for quotation is considered serious, the estimator will first—in his or her mind—compare the product with other jobs of similar products and then search for precedents in personal (or the shop's) records, such as old drawings, book illustrations, or electronic files.

8.2   Precedents

If there are close similarities (precedents), the estimating process is relatively simple, because there is a good basis from which to extrapolate what will be required for the new mold. For example, the precedent can be a mold with only a few cavities for a product with a shape similar to the one for which the mold is to be estimated, for the same number or for more or fewer cavities. In this case it is up to the estimator to find out from records, if possible, how good the mold performed in operation, and if the hours estimated to produce the mold were adequate; in other words, was the customer happy and did the shop make money with this mold? This process is easy if proper records are kept, as was suggested in Chapter 6. The estimator should consult with the people who actually made that mold to find out if there were any problems during manufacture or testing of the mold, and then adjust for it when pricing the mold. With the absence of good records, unfortunately, this is possible only if there was little turnover in personnel in the shop.

8.3   No Precedents

If molds for a similar product have never been made before or the estimator is not familiar with the type of mold requested, there are, in general, two possibilities to be considered.

(1) The estimator will make sketches using previous experience as a mold designer and show at least one method as to how the product could best be made. These sketches will then be the basis for the estimate. (The problem with this method is that it will take much estimating time, and even so, the estimator cannot devote as much time to it as the mold designer will have after the order for this mold has been booked. It is important that any such preliminary sketches are made available to the mold designer, who then may (or may not) follow them for the final design. From the estimator's sketches it is then fairly easy to prepare an estimate. The main problem with this method of estimating is that the estimator makes a bad mistake, typically by not seeing, underestimating, or even ignoring any difficulties that may arise due to a peculiar product shape. The mold designer will then not use these sketches, but will come up with a proper mold design, which could be more (sometimes much more) expensive to build than was first estimated. In this case, any responsible mold maker (whose reputation is at stake) will have no choice but to build this mold, even if it will result in a financial loss. Such losses can then be written off as learning experience or as research and development expenses.

(2) A good alternative is to invite the participation of the client to share in advance the cost of designing the new mold before estimating. This is often very useful if the product is completely new and the projected quantities are extremely large, or where the product is considered very complicated to mold. For a certain quoted price, the mold maker will offer to design either concepts of the mold, or a complete mold. This is also often done for a whole system, that is, not only a mold but including any product handling and postmolding operation of the product. After agreeing with the client that the proposed mold and/or the whole system will do what is needed, the mold and related equipment cost can be fairly easily estimated on the basis of this preliminary design, and there is much less risk of too low or too high an estimate. Traditionally, mold makers add an often quite high safety factor when quoting unfamiliar molds, to cover the unexpected. If the mold is fully designed, there is no need for such insurance; this will result in a lower mold and system price, which benefits the client. The cost of the design paid in advance is then considered in the final mold price. If the client decides not to proceed with the project, at least the mold maker will have the sometimes considerable design expenses paid.

8.4   Methods of Estimating

(1) One method is to actually break down each and every mold part into its estimated cost: material, the cost of the various machining steps (milling,

turning, grinding, EDM, polishing, etc.), the cost of heat treating and other expenditures for finishing in house or by suppliers, the cost of standard hardware, and the costs of assembling and testing the mold. Include also the cost of any fixtures or special tools required in the manufacture of the mold parts. While some of the costs are usually quite simple to establish from price lists and records, this method expects that the estimator or assistants have intimate knowledge of the machining operations involved and the operating times required for each step. Since molds consist of many different parts, this is obviously a slow, time-consuming process; however, as long as the estimator really knows the business well it can yield quite accurate estimates.

(2) The method used most often is to base an estimate on experience from precedents. If, for example, the mold considered has 8 cavities and there is a suitable precedent of a 4-cavity mold, it is fairly easy to extrapolate, by calculating the cost of the new total number of stacks plus the proportional increase of the cost of the larger mold shoe. Many estimators then add a risk factor, which, depending on the difference from the precedent and the general familiarity with the type of mold, may be anywhere up to 50% (or even more) on top of the estimated cost, depending on the mold maker's practice and policies. It is best if the estimator works from a finished product drawing, with all dimensions, and where all tolerances are shown. There is usually little risk if the same mold has been built before, and much risk if there are many unknowns. This method is good if there are good records of many similar molds made over the years; there is less risk of repeating earlier mistakes.

(3) "Ballparking" should be used with care. It requires real experience and solid background in mold making. It should also have the proviso that the quoted price is only a rough estimate and must be confirmed at a later date when all data are ready (including tolerances) and after the order is received.

8.5   Mold Cost and Mold Price

The estimator, in essence, prepares only the foreseen cost to be incurred when building the mold. The cost is the basis for quoting the actual price to the customer. There will be a standard markup on top of the estimated cost, in percentage over the cost, or whatever the company's policy is to cover overhead, expenses, risk (with this mold), and profit. Since every mold is different in size, number of cavities, complexity, and so on, it is usually difficult to create a standard price list for molds, except if many identical molds based on standard mold components are built on a regular basis.

There is another management consideration: The plastics mold business is traditionally up and down, seasonally. In times of low sales, molds may be quoted at prices lower than the costs determined by the estimator, solely to get the job, and to keep the shop busy to avoid layoffs. One unfortunate result of this method is that as soon as the shop is filled with such money-losing molds, as the business picks up again, well-paying jobs may have to wait because the shop is busy.

9 Machining, Mold Materials, and Heat Treatment

9.1 Machining of Mold Components

This section is not meant to be a guide for the actual machining operations, but gives some descriptions of the evolution of machining in mold making. Earliest mold components and plates were produced by first sawing the raw blanks from steel plates of the appropriate thickness bought from the steel mill, with reciprocating or (endless) band saws. The next step was then squaring and/or rough machining these blanks—mostly plates, but also blanks for cavities—on shapers, with the blank held solidly and a single cutting tool moving back and forth over the surface. This slow method was abandoned in favor of rough grinding with special, large grinding machines or milling with large cutting heads in vertical or horizontal milling machines. Both these methods are now used extensively. Since this requires large, expensive machines, which smaller mold makers cannot usually justify economically, a service industry developed, specializing in the machining of the—often large—plates; this was the origin of the mold supply houses such as DME, National, Hasco, and others. While they made (and still make) any requested size within a certain range, the biggest advance came with the standardization of sizes (length, width, thickness), which permitted listing them in catalogues, available for fast delivery. These supply houses rapidly widened their lines by adding other items that, up to then, the molders had made themselves, such as leader pins and bushings, ejector pins and sleeves, and many other mold components and accessories and hardware that have come into use as the industry expanded.

By standardizing designs of these hardware items it became possible to mass produce such parts, using the best-suited materials and finishes, often using specialized machines, thereby making parts not only of better quality, but also at a much lower cost than would be possible in most mold shops, with their limited equipment. Today hardly any mold maker makes mold hardware, but it took quite a while for some to realize the advantages of the quality, the ready availability, and the low cost of these mass-produced parts. (Standard sized screws and nuts have been used for many years.) The supply houses also provide a service to machine large cutouts and openings in standard plates (and mold sets) and the bores for leader pins and bushings to their own standard or the customer's specification, which is very convenient if the mold maker lacks the large and accurate machines to do it in-house.

For the manufacture of the large mold parts, mostly plates, and large cavity blocks, there have also been significant changes. Blank plates are usually purchased, ready rough ground, flat and square to standard or special sizes. They should be somewhat thicker (maybe 0.1 mm, depending on the size of the plate) than the final dimension, to permit regrinding to the final size, if necessary. This is especially important after roughing out large volumes of steel, which may release stresses in the plates, which can result in warpage. The plates (or large cavity blocks) are machined with common machine tools such as lathes, drilling and milling machines, and jig bores. These machines have also improved over the years, becoming much more rigid, allowing the use of better cutting tools, multiple cutting heads, carbide cutters, higher cutting speeds, and the introduction of computerized, numerical controls (CNC). This last advance became possible only after the mold designs improved and began to provide mold part detail drawings. This also necessitated another manufacturing step, the introduction of specialists (production planners or engineers), usually persons with all around experience in machining, to prepare the logical steps in the machining of the parts, that is, the sequence of operations and the tools to use. Up to then, this was usually left to the machinist operating the machine tools; in fact, the old but still widely used practice was that the machinists move with the work pieces from one machine tool to the next until the mold part is finished.

The next step in modernization was to provide the milling machines and so on with automatic tool changers. The responsibility of the machinists became mainly the mounting of the work pieces in the machine tool, seeing that all tools are prepared as specified before installing them in the tool changer, and generally observing the machine to prevent trouble. This gradually eliminated the need for the operator to actually work "hands on" during the cutting process, and even allowed the use of one operator for more than one machine tool. The setup of the work piece in the machine is always critical to ensure the proper reference to specified edges or tooling holes of the work piece. Some of the modern machines don't even require this step in the setup: the machine first feels (reads) the position of the work piece as it is mounted in a jig or fixture, and then automatically adjusts all coordinates to this position.

For smaller stack parts, blanks are still cut from bars or rods and machined with machine tools such as lathes, drilling and milling machines, and jig bores. These machines, too, have improved over the years, by becoming more rigid, using better cutting tools, carbide cutters, higher cutting speeds, and the introduction of CNC. The finish surface and cylindrical grinding of these parts (where required) have also greatly improved over the years with higher cutting speeds and by profile grinding odd flat or round shapes.

Electrical Discharge Machining (EDM) and later the *wire EDM* have been major advances; both permit the shaping or cutting of odd or otherwise difficult-to-machine (or even "impossible") shapes. A main disadvantage of these methods is that they are very slow, that is, they remove much less steel than any chip-removing machine tool, in any given time. They should be used only if there is no other way to cut a shape. These machines usually run automatically. The cutting electrodes for EDM are made from special copper alloys, or from some carbon/graphite composition. They are machined on conventional machine tools; the problem is that they are wearing and getting smaller during operation, and two or more electrodes are required from roughing to final sizing. There are methods of reducing the cost of machining these electrodes, for example, by casting or molding them to shape; these castings are made to order by specialists in this field. Note that the finished surface created by EDM depends largely on the amount of steel removal. (The higher the current through the electrodes, the faster will be the cutting speed.) To produce a fine finish, the operation becomes very slow. But even a rough EDM finish is often good enough for some molding surfaces; the type of finish must always be specified. EDM can be used regardless of the hardness of a work piece; the very first EDM machines were used mainly to remove broken, very hard tools, such as drills and taps, from a work piece. Since the EDM process may take many hours, one operator can usually look after a number of machines.

The increasing specialization of the mold-making business had another impact on the machining methods used. As a result of specialization, and an increase in demand for multicavity molds, the quantity of similar, often standardized components (cavities, cores, stripper rings, etc.) can become so large as to make it possible to introduce fully automatic machines such as automatic (CNC) lathes combined with other operations, with automatic tool changing; such parts can now be produced from steel blanks or rods right up to their prefinished shape (turning inside and outside, drilling, milling, tapping, etc.) ready for the next step, such as heat treatment, in minutes rather than hours, for each part.

Jig grinding, another machining operation, provides precision grinding, both for location and size, of holes (dowels, etc.), or even nonround shapes in cavity work, from diameters as small as 2 mm (0.060 inch). Large, cylindrical shapes can be honed, a practice used for many years in accurate machine building. Both these operations require special, expensive equipment and are often subcontracted to specialists.

Deep hole drilling, also called gun drilling, is a relatively late addition to the mold-making business; it originated in the manufacture of long bores in guns. Special drill bits are used either in attachments to lathes or in special deep hole drilling machines. The process allows the drilling of straight, very deep holes, without the problems of "wandering" encountered with the conventional twist drills. The cutting face of the drill bit is lubricated with pressurized coolant through the center of the bit, and the chips are flushed out along the outside of the bit. (See also *ME*, Chapter 22.) Holes as long as 2 m or even longer, and diameters as small as 8 mm can be drilled fast, and without any significant deviations from the intended straight path. This is of particular importance when drilling cooling and pressure–air channels in large plates, but it is also important for much shorter and smaller-diameter holes often required in cavity and core cooling circuits or in side cores.

Polishing is an important phase in the mold making process. Traditionally, it is done by hand, which is a long, tedious, and therefore expensive process. In the early days, polishing was sometimes farmed out, usually low-paid women who did the polishing at home. Later on, the specifications for polishing were closely scrutinized: is it really necessary to polish this surface? Is the finish as it comes from the milling machine or grinder good enough? Some plastics require good polish, others not. Often, only some areas need good or even exceptional polish, for appearance or for the intended use of the product. By being critical, much time can been saved in this operation. It is up to the mold designer to specify where and how fine to polish. Some of the polishing operations that used to be done manually are now done by automatic machines; the operator mounts the work piece and the machine does the rest. Other, handheld machines, do the reciprocating motion required for the polishing stone or diamond paste. Flat faces are very difficult to polish while maintaining their true flatness, which is especially important if the product requires optical clarity without refraction. In such cases, the use of lapping equipment may be required, in-house or at a specialist, and the mold must then be designed so that the flat surface (of the cavity or core) can be accessed when using a lapping machine. This usually means providing the cavity or core with inserts that can be easily lapped.

Hobbing is another method of making small cavities, such as for bottle caps or other, often odd-shaped forms. A male punch in the shape of the outside of the molded product (including shrinkage allowance for the plastic) is pushed with great force into a soft steel blank. Obviously, the punch must be very hard and strong; the force is on the order of thousands of tons. The making of the hobs and the actual hobbing is done by hobbing specialists, who have the necessary skills and equipment. Around the middle of the 20th century, and later, it was quite common to use this method for multicavity molds. The main advantage is that one punch, while difficult to make, can be used for as many cavities as 30–60, which are then all identical. If, for example, the product has a difficult shape, ornamental ribs or embossings, and even lettering and escutcheons, it is easier to do it once on the outside of a male part, rather than inside of many cavities. Also, the polish on the hob is always perfectly reproduced. If the hob is well polished, so will be the cavities made from it, and will not require additional polishing. By necessity, the steel of the blank must be soft enough to permit the process; but this is too soft to serve for a high-production mold. The blank, after hobbing, must be rough machined on the outside and then carburized and hardened. Because the steel will slightly grow and possibly move in the hardening process, the hard blank must then be ground to fit the bores in the cavity plate. These are all long and expensive operations; with better machining methods, and especially with the advent of EDM, where the final shape of the product can be easily created in the already hardened but otherwise finished cavity blank, the hobbing process is rarely used today.

Electroforming is another method of making cavities, usually for small and long shapes, such as fountain pen barrels. A mandrel of the shape of the cavity wall acts as the electrode in a nickel electrolyte bath. Nickel is slowly deposited on the mandrel until it reaches a desired thickness of about 2 to 3 mm (0.080 to 0.120 inch); the blank is then stripped off the mandrel. The finish of the cavity wall is an exact replica of the finish of the mandrel, so no further polishing is required. The blank must then be machined on the outside to fit a cavity retainer. This method is best done by specialists in this field. It is slow and quite expensive, but sometimes the only way to produce a cavity.

Computerized molecular build-up is a new electrochemical approach to building small, intricate cavities or cores. A computer reads the mold part drawings three dimensionally and builds up, layer by layer, the molecules of the desired mold material until the complete shape is created. This process is still in development and the author knows of no actual molds built, to date.

9.2   Materials Selection

Production molds are almost always made from steel, both for the mold shoe and the cavities, except for certain mold parts where requirements for better heat conductivity suggests the use of beryllium–copper alloys. Sometimes bronzes or even rigid plastics are used for cams and areas where moving parts cannot be lubricated, or must not be for sanitary reasons. Experimental molds may use softer materials such as aluminum, copper alloys, or even special, metal-filled epoxy-type mixtures, or other materials; they will not be discussed further.

The types of steels used depend on the requirements for each application. Throughout this text it has often been said that the designer must always keep costs and planned productivity of the mold in mind. This will often determine the selection of the right steel for each mold component. There are two points to be aware of: (1) the material (mostly steel) represents about 10–15% of the total mold cost and (2) steel costs vary widely, depending on the annual requirement of the mold maker—the higher the requirements, or at least, the more volume is contracted to purchase over a certain period, typically, one year, the lower the base cost of the steel. Also, the blank size has significant bearing on the cost. Per unit of mass, large pieces are cheaper than small ones. In addition, there are weight, cutting, and other charges, so that even though the base price may appear to be low, by the time the piece is cut and delivered, the price is much higher. It is always worthwhile to contact a steel sales person and get all the details about steel pricing. In general, steels, particularly mold and tool steels, are sold by brand names, different for each steel mill, but it is better to specify steels by their generic names or numbers. In general, there is little or no difference between steels of the same specification originating from different suppliers. However, new mold steels are constantly developed for "better" characteristics, and it may become necessary to reevaluate and update the lists of steels used by the designers from time to time.

9.2.1   Steel Properties

Earlier molds used mild steels even for stack parts, but they did not last for long production runs. Mold makers were gradually switching to the types of hardened steels that were used in tool and die making; however, these steels were often too brittle or otherwise unsuitable for mold applications, so the steel industry began to develop steels specifically designed for the plastics industry. The important features for mold makers are essentially as shown below.

(1) *Tensile (or compressive) strength.* This is important for long life of the components as they are subjected to high stresses within the mold, particularly those created by high injection pressures and large clamping forces.Tensile strength Compressive strength

(2) *Toughness*. This is especially important for long, slender cores and inserts subjected to side forces deflecting them during injection.

(3) *Wear resistance*. Wear results from plastic abrasion during injection, and mostly wear from mold parts rubbing against each other.

(4) *Hot hardness*. This is of special importance for hot runner components, but also for molds for some plastics that are molded at high temperatures. Note that many hardened tool steels start to anneal at temperatures lower than the melt temperatures sometimes required for injection.

(5) *Corrosion resistance*. Some plastics attack (corrode) steels and other mold materials. In a high-humidity environment, molds corrode (rust) because of the high water content in the air. In all such circumstances, the mold parts and plates should be chrome or nickel plated, which can be quite expensive, especially when considering the often high handling costs where the plating is performed by outside suppliers. These stack parts and plates can also be made from stainless steel, which is more expensive than other steels, but—considering the overall cost connected with chrome plating—the difference may not be that big, especially if the stainless steels can be bought in large volume.Chromie plated Nickel plated

(6) *Thermal conductivity*. This can be important with high-speed molds. However, keep in mind that, in many cases, good cooling can also be achieved with steel by using a better layout of the cooling channels, and thus avoiding the use of the softer copper alloys, which require more upkeep than steel.

(7) *Ease of hobbing*. See Section 9.1 about hobbing. Hobbing is not much used today.

(8) *Ease of machining*. This is an important consideration. The addition of certain alloying elements to the steel makes it much easier to cut chips; this can make a big difference in the time and the cost of machining.

(9) *Ease of polishing*. Some steels are not well suited for polishing and will not permit or maintain the high surface polish often required. Don't forget, however, that high polish is often not required.

(10) *Ease of nitriding*. Nitriding is a surface treatment applied on top of an already well-hardened and otherwise finished part to provide a very hard surface. It is used mostly to improve the wear characteristic of the steel. To nitride on top of a soft base does not make any sense: the hard (nitrided) surface will collapse under any heavy load because the supporting steel is soft.

(11) *Ease of welding.* In some cases, it may be important to be able to repair a worn mold part by welding. While this, in general, is not a good practice and should be done only in exceptional cases, it may permit a "quick fix" to keep a mold running until it can be properly repaired.

(12) *Cost.* We stated earlier that material constitutes a substantial portion of the mold cost, but cost alone must never be the reason to select any steel. There is only one goal for the mold maker and designer: to produce the best mold for the specified purpose, that is, the mold that will produce the lowest cost of the product for the specified production requirements.

Tables 9.1 and 9.2 are intended to give the designer an overview of some common mold materials. The data are approximate, and may vary somewhat from one manufacturer to another. More about molds steels and application examples can be found in *ME*, Chapter 16.

Table 9.2 shows the average of some of the properties of the above materials that are of interest to the mold designer.

By studying the various steels, it can be seen that all steels have only a few of the characteristics required for a certain purpose; typically, a steel may be

Table 9.1   Comparison Chart of a Few Selected Mold Materials

Item	Type	AISI Designation	DIN Material No.	Steel Code	Recommended Hardness (Rc)
1	Prehardened	4140	1.7225	42CrMo4	30–35
2		P20	1.2330	40CrMnMo7	30–35
3	Stainless steel Prehardened	420SS	1.2083	X42Cr13	30–35
4	Carburizing steels	P5			59–61
5		P6	1.2735		58–60
6	Oil hardening	O1	1.2510	106WCr6	58–62
7	Air hardening	H13	1.2344	X40CrMoV5 1	49–51
8		A2	1.2363	X100CrMoV5 1	56–60
9		D2	1.2379	X155CrVMo12 1	56–58
10	Stainless steel (SS)	420SS	1.2083	X42Cr13	50–52
11	High-speed	M2	1.3343	S-6-5-2	60–62
12	Beryllium–copper	BeCu			28–32[a]

[a] It is customary to indicate hardness of machinery steels and bronzes in the Brinell scale. The above chart, however, uses equivalent Rockwell "C" values to give a better comparison with the hardness of tool steels.

Table 9.2　Comparison of the Properties of Different Mold Materials

Item	Wear Resistance	Toughness	Compressive Strength	Hot Hardness	Corrosion Resistance	Thermal Conductivity	Hobbability	Machinability	Polishability	Nitriding ability	Weldability
1	F	VG	F	F	P	G	P	G	G	F	F
2	F	E	F	F	F	G	P	G	VG	G	F
3	F	E	F	F	G	F	P	F	E	VG	F
4	VG	G	G	G	F	F	E	E	VG	VG	E
5	VG	VG	G	G	F	F	VG	E	VG	VG	VG
6	VG	F	E	G	P	G	G	VG	VG	F	F
7	G	VG	VG	VG	F	F	G	E	VG	E	G
8	E	F	E	VG	F	F	F	VG	G	VG	F
9	E	F	VG	VG	F	F	F	F	E	E	P
10	G	G	G	VG	VG	F	F	VG	G	VG	G
11	E	P	E	E	F	F	F	F	E	E	F
12	F	P	P	F	G	E	E	E	E	N/A	VG

Note. Item numbers 1–12 refer to the material types in Table 9.1. P, poor; F, fair; G, good; VG, very good; E, excellent.

very tough but not be very hard or it may not readily accept nitriding. Therefore, the designer will always have to find the most suitable compromise when selecting a steel for a mold part. Some very expensive steels are occasionally used in molds: tungsten carbides are very hard and three times as stiff as steel, but also very brittle and difficult to produce mold parts; and "maraging" steels are tough, hard, and very stable steels that do not move in the hardening process. New steels are constantly being developed by steel manufacturers, with better properties than before, for general mold making and for new applications, to keep pace with the development of new plastics and with new methods of use in mold making.

9.3   Heat Treatment

We will not go into details of the metallurgy and the behavior of steels during heat treatment and the various hardening methods. Basically, the steel structure of certain steels can be changed by heating and subsequent chilling of the work piece to increase the hardness of the steel from a hardness (usually soft) suitable for machining to the hardness that will provide good working life of the steel under repeated exposure to heat, high pressure, and wear. In general, only steels with a carbon content of at least 0.35% can be hardened. So-called mild steels, with lower carbon content (usually in the range of 0.1–0.3%), cannot be hardened. However, to use these relatively inexpensive steels for mold parts that need good hardness, the surface of such mild steels can—after machining to their shape—be provided with a carbon-rich skin by the process of "carburizing," that is, subjecting the work piece at high heat, for about 24 to 48 hours, to a carbon-rich atmosphere. This causes the surface to absorb carbon to a depth of usually between 0.5 and 1.5 mm (0.020 and 0.060 inch). The work piece can then be hardened like through-hardening tool steels, by heating to a high-temperature, quenching in water or oil, and then tempering (reheating to a lower temperature than before quenching) and finally cooling in air.

At the beginning of the "plastics revolution," most molds were made from these mild steels, and special alloys were designed to provide better polishability. The *advantage* of these steels is their relatively low cost, ease of machining, and availability. The *disadvantages* are the costs for carburizing and the subsequent heat treatment: during carburizing, the steel often distorts and even grows slightly. The art in using these steels is to foresee such changes and to allow enough material for grinding after heat treatment to arrive at the final mold dimensions. Since the carbon content diminishes with its depth, which is dependent on the time required for carburizing, the danger is that too much grinding allowance can make the hardened skin disappear during grinding, and in such areas the surface hardness is then as soft as the base steel. Tool and mold steels are "through-hardened," and the amount of grinding to size will not affect the surface hardness. For these reasons, over the years, mild mold steels have been used less and less.

Today, with the development of better machining methods and more rugged machine tools, larger mold stack parts are made mostly from prehardened mold steels, which are supplied from the steel mills and the supply houses at a hardness of about Rc 30–33. This allows the finish machining of most parts without the need for any heat treatment after machining. Note that very large mold parts may require three steps for finishing: (1) premachining to remove the bulk of the outside and any large openings or cutouts, which may cause stresses within the steel to distort the work piece; (2) the piece should then be stress relieved, before (3) finish machining to the desired close dimensions. Smaller mold stack parts are also often made from prehardened blanks, or, for high production molds, from through-hardened mold or tool steels. After hardening, it may still be necessary to grind or otherwise machine the work piece to the final shape before polishing. Heat treatment is usually done by specialists, thus requiring shipping of the parts to and from them, adding time and cost to the heat treatment. By standardizing a small number of different mold steels requiring hardening, costs can be reduced; for larger mold makers, this may make it economical to provide in-house heat treatment.

Appendix 1 CAD/CAM (Computer-Assisted Design–Computer-Assisted Manufacturing)

As stated before, this book is not about the actual technique of designing (delineating) molds, but about the logic and reasons behind a successful mold design and the questions the designer must consider and answer at every step of the design process. Computers now play an important part in this process, especially if there are many precedents accessible to the designer to be used for new designs and if there is a large collection of standards that can be accessed from computer memories without the need for tediously drawing and redrawing, from simple parts to complicated subassemblies. Also, by using special programs, many calculations can be performed rapidly and accurately, and newly created mold designs can be easily checked for efficiency of plastics flow, cooling, strength of materials, cam motions, and so on.

Because there are so many design programs, the designer usually starts by redrawing the customer's information, which may have been submitted as hard copy (prints) or electronically, but originating from a different system than the one used by the designer. Once the to be molded part (the product) is drawn and dimensioned to the designer's shop rules, a program will be used to add the mold shrinkage to established rules. A constant factor may be used for the product, or different shrinkages may be applicable, as explained earlier in this book.

The designer will now go through the motions as explained earlier, either designing "from scratch," or searching the files for a suitable precedent. If a good precedent is found, it can now be merged with the new product drawing and the mold can be designed. Once completed in principle, various programs can be used to check selected areas (plates, cavities, etc.) for physical strength and to check with other programs the expected efficiency of filling the mold cavities, gate location and sizes, runner sizes, the cooling layout, and so on. Note that all results from using these programs depend on the accuracy of the data provided, such as reasonable assumptions as to temperatures, pressures, times, and so on.

Once the mold drawings are finished, they are transferred to the manufacturing group. By using related, compatible CAM programs, which are often developed in-house, and the input of experienced machinist/programmers, the manufacturing group will determine the best tools to use for the selected machine tools and the appropriate tool paths for each mold part for each tool and for each machine tool.

The following is a list of better known and widely used CAD and CAM programs.

CAD/CAM Programs

Autocad, Autodesk Canada Inc. (mostly for PCs)
   90 Allstate Parkway, Suite 201, Markham, ON, Canada L3R 6H3
   905-946-0928
Unigraphics, Unigraphics Solutions
   2550 Matheson Blvd., Mississauga, ON, Canada L4W 4Z1
   905-212-4500
Proengineer, Parametric Technologies Co.
   128 Technology Dr., Waltham, MA 02453, USA
   781-398-5000

Fluid Flow Programs

CADMOULD, Simcon Inc. (moldflow, cooling, shrinking and warpage)
10914 N 39th. St., Suite B-4, Vancouver, WA 98682
888-754-8628
MOLDFLOW, Moldflow Corp. (mold flow, cooling)
   91 Hartwell Ave., Lexington, MA 02421, USA
   781-674-0085
FEMAP Enterprise (fluid flow, all fluids)
   PO 1172, Exton, PA 19341, USA
   610-458-3660

FIDAP, SPRC (fluid flow)
1155 North Service Rd., Suite 11, Oakville, ON, Canada L6M 3E3
905-465-1733
(partner for Fluid Dynamics International, 708-491-0200)

Mechanical Stresses

ANSIS Mechanical Dynamics Ltd.
400 Carlingview Dr. Toronto, ON, Canada M9W 5X9
416-674-2144
AL GOR
150 Beta Dr., Pittsburg, PA 15288, USA
412-967-2700

Index

2-plate molds 14, 57, 87
3-plate molds 58, 87

accessories 97
accumulator 39, 40
– package 39
accuracy 27
adaptor ring 36
air assist 23, 37
air ejection 23, 37
air-operated actuators 39
alignment 27, 64, 73
– features 80
amorphous plastics 26
angle pins 52
appearance 58, 64
assembly drawing 2, 45, 80
assembly drawings 2
assumptions 109
automatic molding 25
auxiliary actuators 79
auxiliary controls 37
availability 76

backing up 52
backup 55
balanced cavity 54
balanced runners 64
ballooning 81
bending 75
beryllium–copper 70, 101
beverage crates 48
Bill of Materials 45, 81
blades 21
blow downs 39
blower 40

bosses 66
breakers 41
bushings 27, 45, 97

CAD 2, 109
CAM 109
carburizing 106
cavity 11
– construction 54
– shape 11
– space 9, 16, 38
– spacing 59
– walls 16
center gated 58, 79
chase 42, 44, 55
chrome plated 103
circuit breakers 42
clamp stroke 37
clamping force 13, 15, 16, 37
CNC 98, 99
cold and hot runner molds 41, 60
cold runner 14, 38, 56, 57, 63, 64
collapsible cores 51
color changes 57, 59
composite cavities 55
composite cones 55
compressibility 26
compression 75
compressive strength 102
computer 5, 19, 108
conductivity 70
containers 83
contamination 40, 58
continuous vents 67
coolant connections 72, 79, 80
cooling 18, 19, 69

– channels   68, 103
– circuit   72
– lines   71
– lines layout   80
– water supply   41
copper alloys   103
core   11, 55
– backing plate   87
– construction   56
– cooling   87, 90
– shift   56, 64
corrosion resistance   103
cost-effective   42
counter bore   45
cross hatching   81
crystalline plastics   26
cup-shaped product   48
cycle times   34

deep hole drilling   100
deep-draw containers   37
deflection   75
degrade   63
design   6
– review   73
– rules   9
detail drawings   45, 80, 98
diamond paste   100
die casting   3
dowels   72, 75
draft   37
– angles   49
drawings   46
drinking cups   83
drooling   59

economics   3
EDM   99
ejection   22, 68, 78
ejector pins   45, 67, 97
ejector plate return   68, 77
ejector stroke   37
electric power   41
electrical discharge machining (EDM)   99
electroforming   101
electroless nickel   43
electronic drafting   5

estimator   92
Euro   36
experimental molds   102
experimental setup   34
expertise   93

family molds   79
fatigue   68, 75, 76
filtered air   40
fitting   2
flash   13, 74
flashing   74
floating mounting   56
floating platen   78
flow capacity   20
forces   75

gate   14
– location   64
– retainer plate   85
– size   63
grain structure   78
grinding allowance   107
guide pins   85
gun drilling   100

handling the mold   31
hardness   76
hardware   76, 97
heat   69
– barrier   89
– controllers   41
– expansion   88
– loss   63, 64
– losses   59
– treatment   97, 106, 107
heaters   41, 59
high shrinking plastics   83
hobbing   100, 103
holes   48, 77
horn pins   52
hose connections   72
hot hardness   103
hot runner   38, 63, 79
– molds   41, 59, 64, 87
– manifold   63
housings   83

hydraulic supply   40

in-house testing   2
injection blow molding   22
injection molding machine   7
injection pressure   13, 15, 17, 37, 39, 55, 59,
   74, 75
injection speed   38, 39
injection unit   78
insulated runner   62
insulated runner molds   62
interface points   80
internal threads   50

jig grinding   99

L/t ratio   65
laminar   19
land   16, 66
lapping   100
large production   78
latches   79
leader pins   27, 45, 97
leaking   74
learning experience   94
legal implications   47
locating ring   35
low-cost mold   44
lubrication   77

machine ejector   36
machine nozzle   14
machine platen layout   79
machine specifications   35
machinery steel   43
machining   97, 103
machinists   45
manifold   59, 64
– heaters   41, 64
maximum clamp force   17
melt   38
– flow   66
– temperature   17, 25, 59
– temperatures   103
mold   9
– clamps   36
– cost   4, 92, 95

– designers   1
– drawings   45
– hardware   45
– maker   1
– materials   69, 74, 97
– mounting   36, 80
– plates   74
– price   92, 95
– release agents   25
– shoes   42, 44
– steels   43, 102
– supply houses   97
molding cycle   38, 70
molecular build-up   101
mounting plate   83
moving cavity   85
moving platens   78
multicavity molds   11, 30, 31, 71
multilevel   78
multiple gates   56

nameplate   17
nickel plated   103
nitriding   103
notches   77
notes on drawings   47
nozzle radius   36
number of cavities   10
number of gates   37
number of screws   73

open gates   59
overcaps   58

parallel   20, 84
parting line   12, 51
path of least resistance   21, 66
plain bottom   83
plastic fronts   18
plastic inventory   63
plasticizing capacity   38, 78
plastics flow   64
plate cooling   22
plate deflection   68
plate thickness   68
plate supports   80
platen size   35

plates   76, 97
polishing   100, 103
polishing stone   100
postmolding operations   3
power consumption   41
power failure   41
precedents   2, 93, 108
preferred number   11
prehardened mold steels   107
preload   29, 49, 52, 55, 73, 79, 91
press fits   74
pressure air   39
pressure drop   16, 63
product design   33
product drawing   95
production   34, 75, 87
productivity   18, 37, 42, 68, 69
projected area   15, 37, 78
projections   48
protection of the cores   28

quenching   106
quick mold change   36
quotation   92

reentrant bottom   83, 84
records   93
regrinding   58, 60
requirements   34
residence time   25, 60
retractable core   51
reversed flow   65
ribs   50, 66
risk factor   95
robots   23, 75
rollers   52
runner   14
– mold   80
– systems   56, 63

safety factor   77, 94
safety gate   25
screw caps   58
screws   45, 72, 80
self-cleaning   68
self-degating   57, 58
sequence of operation   47

serial   20
shallow engraving   49
shot capacity   38, 78
shot size   37
shrinkage   11, 26, 108
shrinking   70
shut height   36, 78
side cores   3, 13, 37, 48, 49, 52, 55, 90
side forces   13
side wall   48
significant cross section   46, 48
single-cavity molds   56
single-level   78
sink marks   21
sketches   94
sketching   5
sleeves   21, 45, 67, 97
slender core pins   21
slender cores   56, 103
slots   50
snap   49
specialists   4, 107
specialization   93
SPI   36
split cavities   49, 52
split molds   13
splits   90
spot vents   67
springs   78
sprue   14, 64
– bushing   35, 59
– extension   79
stack   42, 55, 71, 80
– layout   48
– molds   78
stainless steels   43, 103
standard hardware   2
standard mold components   95
standard mold shoes   43
standardizing   107
standards   108
start-up   72
stationary platen   78
steel sizes   34
steels   74, 101
strength of material   76
stress relieved   107

stripper plate   23, 87
stripper rings   23, 68, 74, 75, 87
stripping   50
stroke   78
support pillars   68, 84
surface definition   17, 18, 64
surface finish   77
symmetry   10

taper lock   27, 29, 30
taper pins   27, 31
tapers   74
technical products   87
tensile strength   102
tension   75
test machines   2
test report   2, 82
thermal conductivity   103
thermocouples   42
thin-wall molding   77
thin-walled   56, 71
tie bar   79
– clearances   35
title block   81
tolerances   2, 27, 33, 34, 95
torsion   75
toughness   76, 103
T-slots   84
turbulent   19, 71
two-stage ejection   51

undercuts   49, 50
universal mold shoes   44
unscrewing   50, 75

valved gates   59
variable shrinkage   26
vent   65
– channels   66
– gap   66
– grooves   66
– pins   67
venting   18, 65
vertical split   48
views   46, 80
virgin plastics   60
viscosity   16, 17

wall thickness   17
wear   30, 75, 77, 103
wedge   30, 74, 76
– action   75
– locks   27
weld lines   66
welding   104
wet area   72
wire EDM   99

yield strength   77